VAP

Lectures on electromagnetic theory

Lectures on electromagnetic theory

BY

L. SOLYMAR

OXFORD NEW YORK
OXFORD UNIVERSITY PRESS

Oxford University Press, Walton Street, Oxford OX2 6DP
London New York Toronto
Delhi Bombay Calcutta Madras Karachi
Kuala Lumpur Singapore Hong Kong Tokyo
Nairobi Dar es Salaam Cape Town
Melbourne Auckland
and associated companies in
Beirut Berlin Ibadan Mexico City Nicosia

Oxford is a trade mark of Oxford University Press

Revised edition published 1984
Reprinted 1984

British Library Cataloguing in Publication Data
Solymar, L.
Lectures on electromagnetic theory.
1. Electromagnetic theory
I. Title
530.1'41'02462 QC670
ISBN 0–19–856169–5

Printed in Great Britain by
J. W. Arrowsmith Ltd., Bristol

TO

PROFESSOR KÁROLY SIMONYI

who first introduced me to the secrets of electromagnetic theory

Preface

THIS book is an expanded version of my lectures given in Michaelmas term 1974 to second year undergraduates in the Department of Engineering Science, Oxford University. I aimed at a text which is readable, covers a wide range from electrostatics to relativity in a short space, and can still offer a quantitative mathematical treatment.

A feature of the book, not usually found in textbooks, is the inclusion of a large number of quotations from the pioneers of electromagnetism. I have always felt that our courses give too little historical insight. I made here an attempt to whet some appetites for delving into the past. In particular, by giving quotations from Young to Keller, I followed through the story of edge diffraction which I hope others will find as fascinating as I do. In connection with the translations I wish to acknowledge here the help I received from Professor J. L. Ackrill, Mr. R. S. Lucas, Dr. P. C. H. Wernberg-Moller, and Mr. R. P. Aizlewood in the Latin, German, Danish, and Russian languages respectively.

I am greatly indebted to Drs. J. E. Allen, D. M. S. Bagguley, and A. M. Howatson for many interesting and stimulating discussions on the subject of Chapter 4.

For great help in preparing the manuscript I wish to thank my wife Marianne.

I hope many of the mistakes in the first printing have now been eliminated. I am particularly grateful to Dr. N. E. Christensen of the Technical University of Denmark, Lyngby, for sending me a long list of corrections.

Oxford L. S.

Contents

Introduction

The history of electricity is a field full of pleasing objects, according to all the genuine and universal principles of taste, deduced from a knowledge of human nature.

But though other kinds of history may, in some respects, vie with that of philosophy, nothing that comes under the denomination of history can exhibit instances of so fine a rise and improvement in things as we see in the progress of the human mind in philosophical investigations. To whatever height we have arrived in natural science, our beginnings were very low, and our advances have been exceeding gradual. And to look down from the eminence, and to see, and compare all those gradual advances in the ascent, cannot but give the greatest pleasure to those who are seated on the eminence, and who feel all the advantages of their elevated situation. And considering that we ourselves are, by no means, at the top of human science; that the mountain still ascends beyond our sight, and that we are, in fact, not much above the foot of it, a view of the manner in which the ascent has been made cannot but animate us in our attempts to advance still higher, and suggest methods and expedients to assist us in our further progress.

These histories are evidently much more necessary in an advanced state of science, than in the infancy of it. At present philosophical discoveries are so many, and the accounts of them are so dispersed, that it is not in the power of any man to come at the knowledge of all that has been done, as a foundation for his own inquiries. And this circumstance appears to me to have very much retarded the progress of discoveries.

I likewise think myself peculiarly happy in my subject itself. Few branches of natural philosophy would, I think, make so good a subject for a history. Few can boast such a number of discoveries, disposed in so fine a series, all comprised in so short a space of time, and all so recent, the principal actors in the scene being still living.

I entitle the work the *History and present state of electricity*; and whether there be any new editions of the whole work or not, care will be taken to preserve the propriety of the title, by occasionally printing additions, in the same size, as new discoveries are made; which will always be sold at a reasonable price to the purchasers of the book; or given gratis, if the bulk be inconsiderable.

JOSEPH PRIESTLEY *The history and present state of electricity with original experiments* London 1767

THE subject of electricity can be broadly divided into three big branches: circuits, electromagnetism, and electric properties of materials. You know what circuits are like. They consist of lumped elements interconnected in various ways. The elements themselves are simple; they can be represented by boxes with a number of terminals into which currents may flow and over which voltages may exist. It is not difficult to conceive what each box stands for. Nevertheless when hundreds, thousands, or may be millions of boxes are interconnected it is well beyond the capacity of ordinary mortals to comprehend what is going on. I never know whom to admire more, the academic engineer who designed the whole complex apparatus or the industrial engineer who comes and repairs the thing when something goes wrong. Anyway the point I wish to make is that complexity in circuits is primarily due to the large number of components.

If we wish to understand the electric properties of materials the situation is a lot worse. Then even the simplest elements display a rather complex behaviour. Take for example the hydrogen atom; it consists of one proton and of one electron—you cannot have anything simpler than that. If you want to unravel the properties of the hydrogen atom you need to do pages and pages of mathematics. And if you venture as far as the helium atom you will be told by mathematicians that no exact solution of the differential equation exists. Imagine now 10^{22} atoms heaped upon one another, each one containing 50 electrons, and try to answer the question about the electrical conductivity of the material. It is a miracle that such questions can sometimes be answered.

Electromagnetism is somewhere half-way between circuits and electric properties of materials. Some of the phenomena are fairly simple, so simple in fact that they are taught in secondary schools: you must have studied Coulomb's law, Biot–Savart's law, the law of induction, and a few similar laws concerned with charges, forces, currents, and voltages. You must also have heard about Snell's laws of reflection and refraction, about lenses and mirrors, about diffraction effects, etc. All these phenomena were treated separately, making only very faint attempts to relate them to each other. The reason is that it is not easy to connect them. Electromagnetic phenomena have such a wide range, display such varied properties, that explanation from any single angle is bound to fail. One may accept that a compass will always point towards the North. Every child knows that. One may equally accept that a light ray will refract when entering water. This is nothing other than putting personal experience into scientific jargon. But to say that the two things are just different manifestations of the same phenomenon is stretching one's credulity a little too far. There is no one—certainly no one I know—to whom such a connection would seem obvious. The links that exist are not physical, they are purely mathematical. The similarity between the magnetic field created by an electric current and that present in a light beam is not tangible. The sole reason why we may call both of them a 'magnetic field' is that they obey the same set of equations.

The conclusion is inescapable. We are in for a lot of mathematics. Having done the mathematics the situation will start to improve, traces of physical intuition will gradually emerge. And the more mathematics you do, the more robust your physical intuition will become. To my mind the greatest difference between electromagnetism and electric properties of materials is that the former may be understood. Not without a great deal of study, not without some sweat and toil, but eventually one gets there. After some steadfast climbing one breaks through the clouds and arrives at a pinnacle from where the whole mountain range may be clearly seen. Well, some of the peaks may be under some light fog but that is of little consequence. Even if we do not know the exact shape of one particular mountain and have no reports on the weather conditions we do know that it is only up to us to set forth on an expedition and conquer the peak. Electromagnetism has still some uncharted territory but contains no more surprises. I know most physicists would strongly disagree with this view. I am really looking at the problem with the eye of an engineer. If we do not wish to enquire into such questions as the size of the electron or where the mass of the electron is coming from then we are in the clear.

Now what about materials? Can we study electromagnetism without enquiring into the properties of materials? Are not all engineering constructions made of real materials possessing widely different properties? Yes, of course; I have no intention of restricting the study to phenomena occurring in a vacuum. We have to take into account *some* of the properties of materials but we shall not be too inquisitive. In most cases we shall characterize a material by three macroscopic constants: permittivity, permeability, and conductivity. This is again a view many teachers of electromagnetism would disagree with. One can argue that the microscopic properties are not to be ignored, and whenever one has a chance one should discuss the interaction between matter and electromagnetic waves. I feel that electromagnetism as I define it is difficult enough. First try to understand it in that limited context; you can break down the self-imposed barrier later when you will be concerned with properties of materials. For example, in the study of solids and plasmas it is regularly assumed that charged-particle pairs can disappear by recombination. We shall have none of that here; all our particles will be considered indestructible.

A further question I would like to touch upon is the necessary compromise between depth and breadth, rigour and intuition. Traditionally, this is a subject where rigour is highly valued. For a long time it has been the playing ground of mathematically minded engineers and practically minded mathematicians. The trouble with rigour is that besides being a bore it is also very elusive merchandise. However rigorous you might attempt to be it is always possible to find a man to whom it will appear contemptuously sloppy. It is not my intention to decry the importance of rigour. It can certainly have value in the solution of engineering problems. A 'uniqueness theorem', for example, will assure you that having found one solution there is no point searching for another one. But by and large engineers can survive without rigour, and in my opinion this should be reflected in the teaching of the subject as well. Accordingly, I shall depart somewhat from the line followed by many standard introductory texts and will attempt to cover a wider range of phenomena at the expense of mathematical rigour.

1. Maxwell's equations

One of the chief peculiarities of this treatise is the doctrine which asserts, that the true electric current on which the electromagnetic phenomena depend, is not the same thing as the current of conduction, but that the time variation of the electric displacement must be taken into account in estimating the total movement of electricity.

JAMES CLARK MAXWELL *A treatise on electricity and magnetism* Oxford 1873

Auf die Frage „Was ist die Maxwell'sche Theorie?" wüsste ich also keine kürzere und bestimmtere Antwort als diese: Die Maxwell'sche Theorie ist das System der Maxwell'schen Gleichungen.

HEINRICH HERTZ *Untersuchungen über die Ausbreitung der elektrischen Kraft* Leipzig 1894

ALL the problems we shall be concerned with may be solved by calling upon one or more of the following equations:

$$\nabla \times \mathbf{H} = \mathbf{J} + \frac{\partial \mathbf{D}}{\partial t}, \tag{1.1}$$

$$\nabla \times \mathbf{E} = -\frac{\partial \mathbf{B}}{\partial t}, \tag{1.2}$$

$$\nabla \cdot \mathbf{D} = \rho, \tag{1.3}$$

$$\nabla \cdot \mathbf{B} = 0, \tag{1.4}$$

$$\mathbf{D} = \varepsilon \mathbf{E}, \tag{1.5}$$

$$\mathbf{B} = \mu \mathbf{H}, \tag{1.6}$$

$$\mathbf{F} = q(\mathbf{E} + \mathbf{v} \times \mathbf{B}). \tag{1.7}$$

Where do the above equations come from? They are contained (though not quite in the same form and not in the same system of units) in Chapter IX of Maxwell's *Treatise on electricity and magnetism* published in 1873.

Are we to conclude that electromagnetic theory has made no advance in the course of a century? That conclusion would essentially be correct. Our technique of solving the above equations has improved, and of course we are in a much better position now to evaluate the material constants, but fundamentally electromagnetic theory stands now as it stood a century ago.

As far as the interrelationship of electromagnetic quantities is concerned Maxwell knew as much as we do today. He did not actually suggest communication between continents with the aid of geostationary satellites, but if he was taken now to a satellite ground-station he would not be numbed with astonishment. If we would give him half an hour to get over the shock of his resurrection he would quietly sit down with a piece of paper (the back of a bigger envelope, I suppose) and would work out the relevant design formulae.

The lack of advance on our part should not be attributed to the idleness of a century, much rather to the genius of Maxwell. The moment he conceived the idea of the displacement current, a new era started in the history of mankind. Events of similar importance did not occur often. Newton's *Principia* and Einstein's first paper on relativity would qualify, and perhaps two or three more learned papers, but that's about all. If we assume that our kind of beings will still be around a few millennia hence, I feel certain that the nineteenth century will mainly be remembered as the century when Maxwell formulated his equations.

What was so extraordinary about Maxwell's contribution? It was the first (and may be the best) example of reaching a synthesis on the basis of experimental evidence, mathematical intuition, and prophetic insight. The term $\partial \mathbf{D}/\partial t$ in eqn (1.1) had no experimental basis at the time. By adding this new term to the known equations he managed to describe *all* macroscopic electromagnetic phenomena. And when relativity came, Newton's equations were found wanting but not Maxwell's; they needed no relativistic correction.

I could go on for a long time in praise of Maxwell. Unfortunately we have little time for digressions however entertaining they might be. But before we get down to the equations I must say a few words in defence of the approach I choose. I know it must be hard for anyone to accept a set of equations without going through the usual routine of presenting the relevant experimental justifications. It might seem a little unreasonable at first sight but believe me this is a possible approach, and under the circumstances it may very well be the best approach. You are already familiar with the mathematical operations curl, div, and grad (I prefer using them in vector-operator form), and you need no introduction to the concepts of electricity. You have heard about electric charge, current, magnetic field, etc. All that eqns (1.1)–(1.7) do is to give a number of relationships between these quantities. So if some of them are known, you can use the equations to work out some of the others.

The notation is fairly standard but still I should better say what is what: $\mathbf{H}$, magnetic field strength; $\mathbf{E}$, electric field strength; $\mathbf{D}$, electric flux density; $\mathbf{B}$, magnetic flux density; $\mathbf{J}$, current density; ρ, charge density; $\mathbf{F}$, force, q, charge; $\mathbf{v}$, velocity of moving charge; μ, permeability; ε, permittivity. The two latter quantities are constants depending on the material under study. Using the subscript zero to denote their values in free space, they come to

$$\varepsilon_0 = 8{\cdot}854 \times 10^{-12}\ \mathrm{A\,s\,V^{-1}\,m^{-1}}, \qquad \mu_0 = 4\pi \times 10^{-7}\ \mathrm{V\,s\,A^{-1}\,m^{-1}}. \tag{1.8}$$

In any other medium

$$\varepsilon = \varepsilon_0 \varepsilon_r \quad \text{and} \quad \mu = \mu_0 \mu_r, \tag{1.9}$$

where ε_r is the *relative permittivity* (or dielectric constant) and μ_r is usually referred to as the *relative permeability*. Notice that

$$\varepsilon_0 \mu_0 = 1/c^2, \tag{1.10}$$

where c is the velocity of light in free space.†

The rest of the course will be concerned with the various solutions of eqns (1.1)–(1.7). Isn't this boring for an engineer? Shouldn't this be done by mathematicians or by computer programmers? Not for the time being. Perhaps one day computers will be big enough and numerical analysts clever enough so that the engineer will only have to pose the problem, but not yet, and not, I think, for some time to come.

In the large majority of cases a straightforward mathematical solution is just out of the question. So one has to use that delicate substance known as physical intuition. How can one acquire physical intuition? There is no easy way. One has to start with a simple physical configuration, solve the corresponding mathematical problem, then solve a similar problem and then another problem, and then a little more difficult problem, and so on. The first breakthrough comes when one can predict a solution without actually doing the mathematics.

In order to have a unified view of the subject we have started with Maxwell's equations. It means a new approach but not a radical departure. The subject is still the same. You will be able to see that the laws you love and cherish (Coulomb's, Biot–Savart's, Snell's, etc.) all follow from our eqns (1.1)–(1.7).

The order of discussion will follow the traditional one: electrostatics first, followed by steady currents, then we shall move on to slowly varying phenomena, and reach finally the most interesting part, fast-varying phenomena, exhibiting the full beauty of Maxwell's wonderful equations.

† It is no coincidence that the product of ε_0 and μ_0 is related to c^2. Strictly speaking we should have said $\mu_0 = 4\pi \times 10^{-7}$ and should have defined ε_0 as $1/\mu_0 c^2$, where c is obtained from measurement. But why should μ_0 be $4\pi \times 10^{-7}$ and why should it have the dimensions volt × second/ampere × metre? This is a problem intimately related to the choice of units. Since we are using SI (Système Internationale, known previously as MKS by electrical engineers) units and since I hope all other units will go out of fashion, it does not appear to be worthwhile to waste much time upon discussing their interrelationship.

2. Electrostatics

The English philosophers, and perhaps the greater part of foreigners too, have now generally adopted the theory of *positive* and *negative* electricity. As this theory has been extended to almost all the phenomena, and is the most probable of any that have been hitherto proposed to the world, I shall give a pretty full account of it.

According to this theory, all the operations of electricity depend upon one fluid *sui generis*, extremely subtile and elastic, dispersed through the pores of all bodies; by which the particles of it are as strongly attracted, as they are repelled by one another.

When the equilibrium of this fluid in any body is not disturbed; that is, when there is in any body neither more nor less of it than its natural share, or than that quantity which it is capable of retaining by its own attraction, it does not discover itself to our senses by any effect. The action of the rubber upon an electric disturbs this equilibrium, occasioning a deficiency of the fluid in one place, and a redundancy of it in another.

I shall close the account of my experiments with a small set, in which, as well as in the last, I have little to boast besides the honour of following the instructions of Dr. Franklin. He informed me, that he had found cork balls to be wholly unaffected by the electricity of a metal cup, within which they were held; and he desired me to repeat and ascertain the fact, giving me leave to make it public.

Accordingly, December the 21st. I electrified a tin quart vessel, standing upon a stool of baked wood; and observed, that a pair of pith balls, insulated by being fastened to the end of a stick of glass, and hanging entirely within the cup, so that no part of the threads were above the mouth of it, remained just where they were placed, without being in the least affected by the electricity;

May we not infer from this experiment, that the attraction of electricity is subject to the same laws with that of gravitation, and is therefore according to the squares of the distances; since it is easily demonstrated, that were the earth in the form of a shell, a body in the inside of it would not be attracted to one side more than another?

JOSEPH PRIESTLEY *The history and present state of electricity with original experiments* London 1767

Loi fondamentale de l'Électricité.

La force répulsive de deux petits globes électrisés de la méme nature d'électricité, est en raison inverse du carré de la distance du centre des deux globes.

L'électricité des deux balles diminue un peu pendant le temps que dure l'expérience; j'ai éprouvé que, le jour où j'ai fait les essais qui précèdent, les balles électrisées se trouvant par leur répulsion à 30 degrés de distance l'une de l'autre, sous un angle de torsion de 50 degrés, elles se sont rapprochées d'un degré dans trois minutes; mais comme je n'ai employé que deux minutes à faire les trois essais qui précèdent, l'on peut, dans ces expériences, négliger l'erreur qui résulte de la perte de l'électricité. Si l'on desire une plus grande précision, où lorsque l'air est humide, & que l'électricité se perd rapidement, l'on doit, par une première observation, déterminer la doit ou la diminution de l'action électrique des deux balles dans chaque minute, & se servir ensuite de cette première observation, pour corriger les résultats des expériences que l'on voudra faire ce jour-là.

CHARLES AUGUSTIN COULOMB *Premier mémoire sur l'electricité et le magnetisme, Histoire de l'Académie Royale des Sciences* Paris 1785

Dans un Corps conducteur chargé d'Électricité, le fluide électrique se répand sur la surface du corps, mais ne pénètre pas dans l'interieur du corps.

CHARLES AUGUSTIN COULOMB *Quatrième mémoire sur l'électricité Histoire de l'Academie Royale des Sciences* Paris 1786

Lorsqu'une science déja fort avancée a fait un pas important, il s'établit des liaisons nouvelles entre les branches qui la composent: on aime alors à porter ses regards en arrière pour mesurer la carrière qui a été parcourue, et voir comment l'esprit humain l'a franchie. Si nous remontons ainsi à la naissance de l'électricité, nous la trouvons, au commencement du dernier siècle, réduite aux seuls phénomènes d'attraction et de répulsion; Dufay, le premier, reconnut les règles constantes auxquelles ils sont assujettis, et expliqua leurs bizarreries apparentes. Sa découverte des deux électricités, résineuse et vitrée, fonda les bases de la science; et Franklin, en la présentant sous un nouveau point de vue, en fit le fondement de sa théorie, à laquelle tous les phénomènes, même celui de la bouteille de Leyde, vinrent naturellement se plier. Epinus acheva de prouver cette théorie, la perfectionna en l'assujettissant au calcul, et parvint, à l'aide de l'analyse, jusqu'à ces phénomènes que le citoyen Volta a si heureusement employés dans le condensateur et dans l'électrophore. La loi rigoureuse des attractions et des répulsions électriques manquoit encore, elle fut établie par des expériences exactes; et, se liant à celle du magnétisme, elle se trouva la même que pour les attractions célestes. On sait que le citoyen Coulomb est l'auteur de cette découverte.

Enfin parurent les phénomènes galvaniques, si singuliers dans leur marche, et si différens en apparence de tout ce que l'on connoissoit déja. On créa d'abord, pour les expliquer, un fluide particulier; mais par une suite d'expériences ingénieuses, conduites avec sagacité, le citoyen Volta se propose de les ramener à une seule cause, le développement de l'électricité métallique; les fait servir à la construction d'un appareil qui permet d'augmenter à volonté leur force, et les lie, par ses résultats, avec des phénomènes importans de la chimie et de l'économie animale.

Rapport sur les expériences du citoyen Volta, par le citoyen Biot, Memoires de l'Institut National des Sciences et Arts Tome V Fructidor An XII

2.1. Introduction

STATIC means not varying as a function of time. So all our quantities ρ, **J**, **E**, **B**, **D**, and **H** will be independent of time. Is there such a thing as time-independent charge? Yes, there is. It means that neither the magnitude nor the position of the charge varies as a function of time. And similarly we can imagine constant electric and magnetic fields. Can we talk of time-independent current? Well, we have to permit the motion of charges to get any current but if the amount of charge crossing a certain cross-section is always the same then the current at that point is independent of time. On this basis constant currents also belong to the static branch of electricity. It is, though, usual to distinguish between electrostatics and magnetostatics; in the former case the variables are ρ, **E**, and **D**, whereas in the latter case they are **J**, **H**, and **B**.

We shall now proceed with the equations of electrostatics, which may be obtained from eqns (1.1)–(1.7) by substituting $\partial/\partial t = 0$ and assuming that **v**, **J**, **H**, and **B** are all zero. We get then

$$\nabla \times \mathbf{E} = 0 \tag{2.1}$$

$$\nabla \cdot \mathbf{D} = \rho, \tag{2.2}$$

$$\mathbf{D} = \varepsilon \mathbf{E}, \tag{2.3}$$

$$\mathbf{F} = q\mathbf{E}. \tag{2.4}$$

We shall introduce now a scalar function ϕ by the relationship

$$\mathbf{E} = -\nabla\phi. \tag{2.5}$$

The physical significance of this new function may be recognized by determining the work performed by carrying a charge from point a to point b:

$$W = -\int_a^b \mathbf{F} \cdot \mathrm{d}\mathbf{s}, \tag{2.6}$$

where the negative sign is due to the fact that the work is done against the electrical forces. Substituting eqns (2.4) and (2.5) into eqn (2.6) we get

$$W = -q\int_a^b \mathbf{E} \cdot \mathrm{d}\mathbf{s} = q\int_a^b \nabla\phi \cdot \mathrm{d}\mathbf{s} = q\{\phi(b) - \phi(a)\}, \tag{2.7}$$

where $\phi(b)$ and $\phi(a)$ are the values of the function ϕ at the end-points of the path (Fig. 2.1). We have used here a mathematical theorem stating that the line integral of a gradient depends only on the end-points and not on the connecting path. The potential at point b may be written with the aid of eqn (2.7) in the form

$$\phi(b) = \phi(a) - \int_a^b \mathbf{E} \cdot \mathrm{d}\mathbf{s}. \tag{2.8}$$

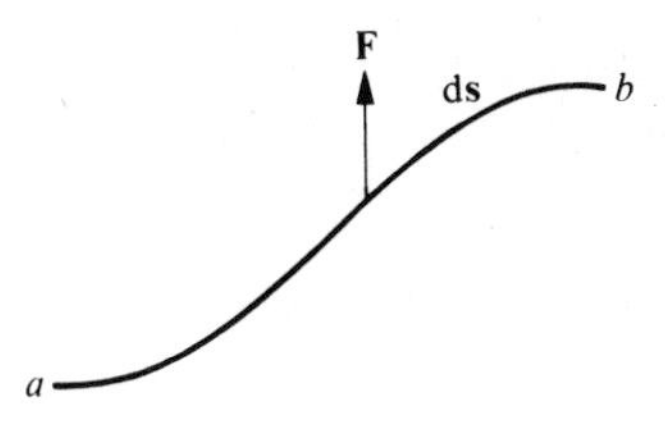

FIG. 2.1. A path between points a and b for determining the work done.

Alternatively, we may choose† $\phi(a)=0$, leading to

$$\phi = -\int_a^b \mathbf{E} \,.\, \mathrm{d}\mathbf{s}. \tag{2.9}$$

What is the good of introducing ϕ?

1. Owing to its scalar character it is more easily calculable than the electric field. Thus very often in practice we determine ϕ first and $\mathbf{E}$ afterwards,
2. The choice of ϕ in the form of eqn (2.5) immediately ensures that eqn (2.1) is satisfied:

$$\nabla \times \nabla\phi \equiv 0, \tag{2.10}$$

so we have less to worry about.

If the potential function is so useful shouldn't we express all our equations in terms of ϕ? Yes, in fact this is what other people did in the past. So let us convert eqn (2.2) as well. Noting that

$$\nabla \,.\, \mathbf{D} = \varepsilon \nabla \,.\, \mathbf{E} = -\varepsilon \nabla \,.\, \nabla\phi = -\varepsilon \nabla^2 \phi, \tag{2.11}$$

we obtain

$$\nabla^2 \phi = -\rho/\varepsilon, \tag{2.12}$$

which is known as Poisson's equation. If $\rho = 0$, the above equation reduces to

$$\nabla^2 \phi = 0. \tag{2.13}$$

This equation is known after another Frenchman as Laplace's equation.

We have not quite finished. There is one more equation often used in electrostatics that contains charge and not charge density. We can get it by integrating eqn (2.2) over a volume τ:

$$\int_\tau \nabla \,.\, \mathbf{D}\, \mathrm{d}\tau = \int_\tau \rho \, \mathrm{d}\tau. \tag{2.14}$$

† The physically measurable quantities like the electric field or the potential difference are independent of the choice of $\phi(a)$.

Using Gauss' theorem for the left-hand side and noting that the volume integral of charge density is just the total amount of charge, eqn (2.14) takes the form

$$\int_S \mathbf{D} \cdot \mathrm{d}\mathbf{S} = q, \tag{2.15}$$

where S is the closed surface of volume τ and q is the charge inside. Eqn (2.15) is known as Gauss's law.

2.2. Coulomb's law

Having got the equations what shall we do with them? Let us first try to prove something that you have come across in school, Coulomb's law. Note the difference. We shall not *postulate* Coulomb's law, we are going to derive it.

We shall have to introduce point charges, but let us be a little more general to begin with and assume that the charge is uniformly distributed within a sphere of radius r_0. We shall now apply Gauss' law and take the surface of integration at a radius $r > r_0$. Notice that everything is spherically symmetric, hence D must be constant on the chosen surface, and the integral comes to

$$\int \mathbf{D} \cdot \mathrm{d}\mathbf{S} = 4\pi r^2 D, \tag{2.16}$$

where $D = |\mathbf{D}|$.

In view of eqns (2.15) and (2.16)

$$D = q/4\pi r^2 \tag{2.17}$$

or, in a vacuum,

$$E = q/4\pi\varepsilon_0 r^2. \tag{2.18}$$

Interestingly, the electric field does not depend on the actual positions of the charges. As long as the charge distribution is spherically symmetric, and as long as all the charges are inside the sphere of radius r_0, the electric field depends only on r and not on r_0. So we can just as well imagine that all the charge is concentrated at the origin of the coordinate system.

Let us place now another bunch of charge (say q_2) into another discrete point a distance r_{12} away from our first charge that we will now denote by q_1. With the aid of eqn (2.4) we get for the force upon q_2,

$$F = q_2 E, \tag{2.19}$$

and substituting for E from eqn (2.18) we obtain

$$F = \frac{1}{4\pi\varepsilon_0} \frac{q_1 q_2}{r_{12}^2}. \tag{2.20}$$

In words:

> the force between two charges at rest is proportional to the product of the charges and inversely proportional to the square of the distance between them. The direction of the force is obvious, it can only act in the line connecting the two point charges.

So we have got Coulomb's law.

2.3. The potential due to charges

The potential due to a point charge may be determined with the aid of eqns (2.8) and (2.18), as follows:

$$\phi(b)=\phi(a)-\int_a^b \frac{q}{4\pi\varepsilon_0}\frac{\mathrm{d}r}{r^2}$$

$$=\phi(a)+\frac{q}{4\pi\varepsilon_0}\left(\frac{1}{r_b}-\frac{1}{r_a}\right). \qquad (2.21)$$

The usual convention is to choose the reference point at infinity and choose the corresponding potential $\phi(a)=0$, so we get for the potential of a point charge

$$\phi=\frac{q}{4\pi\varepsilon_0 r}. \qquad (2.22)$$

If we have a number of point charges $q_1, q_2 \ldots, q_n$ at distances $r_1, r_2 \ldots, r_n$ from the point where we wish to evaluate the potential then we can simply add all the potentials, leading to

$$\phi=\frac{1}{4\pi\varepsilon_0}\sum_{i=1}^{n}\frac{q_i}{r_i}. \qquad (2.23)$$

If instead of point charges we have a distributed space charge $\rho(x', y', z')$ then the sum in eqn (2.23) goes over into an integral:

$$\phi(x, y, z)=\frac{1}{4\pi\varepsilon_0}\int_\tau \frac{\rho(x', y', z')\mathrm{d}\tau}{r}, \qquad (2.24)$$

where r is now the distance between the elementary charge $\rho\,\mathrm{d}\tau$ located at point (x', y', z') and the point (x, y, z) where the potential is evaluated.†

The formulae look quite reasonable; what you need is a little practice in handling them. But as I said before the game is not a purely mathematical one; it is a mixture of physics and mathematics, a combination of intuition and technique. So before we embark upon solving concrete examples let us turn to a graphical illustration of the electric field.

† Note that in the case when there are some further charges *outside* τ their effect must be accounted for by integrals over the closed surface of volume τ.

We shall introduce field lines defined by the statement that at each point on the line the tangent is in the direction of the electric field. The magnitude of the electric field may be represented at the same time by the density of field lines. The nearer they are to each other the greater is the field.

A particularly simple example is provided by a point charge. The electric field is always in the radial direction, so the field lines are just straight lines as shown in Fig. 2.2. According to convention the arrows on the lines point outwards from a positive point charge.

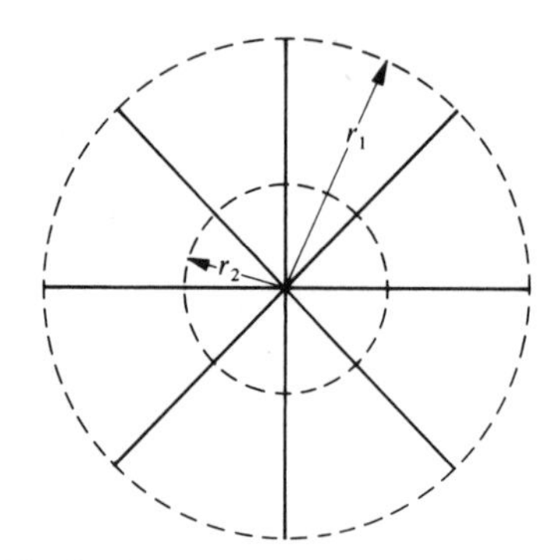

FIG. 2.2. The field lines and equipotentials of a point charge.

Owing to radial symmetry the potential is constant on a spherical surface at a distance r from the point charge. Some of these equipotential surfaces are shown in Fig. 2.2 with dotted lines. Notice that the field lines are perpendicular to the equipotential surfaces. This is of course no coïncidence. The gradient of a scalar function ϕ is a vector perpendicular to the $\phi = $ constant surface and points in the direction of increasing values of ϕ. Since

$$\phi = \frac{q}{4\pi\varepsilon_0 r} \tag{2.22}$$

and in the present case $q > 0$, we find that ϕ decreases with increasing r. Hence the vector $\nabla\phi$ points inwards and $-\nabla\phi$ points outwards. So the whole picture is consistent.

Having completed the calculations for the potential of a point charge, we shall now investigate a more complicated situation. Yes, you guessed correctly, we are going to investigate the equipotential surfaces and field lines of *two* point charges.

In view of eqn (2.23) the potential due to two point charges may be written as

$$\phi = \frac{1}{4\pi\varepsilon_0}\left(\frac{q_1}{r_1} + \frac{q_2}{r_2}\right), \tag{2.25}$$

which for the coordinate system of Fig. 2.3 reduces to

$$\phi = \frac{1}{4\pi\varepsilon_0}\left[\frac{q_1}{\{x^2+(y-d/2)^2+z^2\}^{\frac{1}{2}}}+\frac{q_2}{\{x^2+(y+d/2)^2+z^2\}^{\frac{1}{2}}}\right]. \quad (2.26)$$

The equipotential surfaces may now be plotted on the basis of the above equation and then the field lines may be obtained as trajectories orthogonal

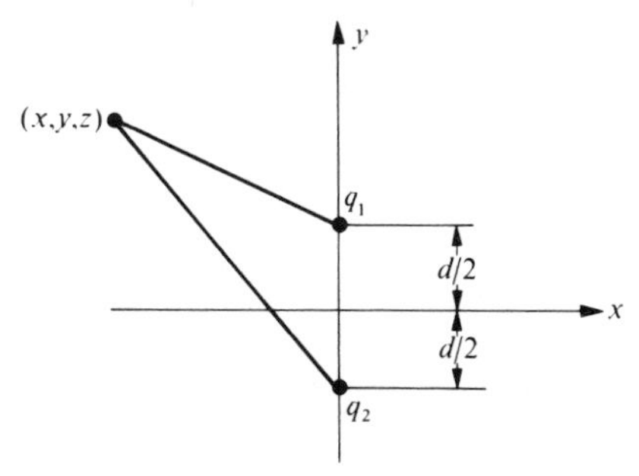

FIG. 2.3. Two point charges.

to the equipotentials. This is simple in principle but extremely messy and tedious if you want to solve it analytically. The best method nowadays, of course, is to put it on the computer and let the computer plot the lot. For $q_2 = -q_1$ the plot in the x, y plane is shown in Fig. 2.4.

If you look at the Figure and study it carefully you will be able to recognize a number of 'commonsense' points. It is common sense once you see the solution but it is unlikely that you would have thought of all of them *a priori*. For example,

1. all the field lines originate on the positive charge and terminate on the negative charge;
2. the potential is zero on the $y=0$ plane;
3. one of the field lines coincides with part of the y axis connecting the two charges.
4. The $x=0$ and $y=0$ planes are planes of symmetry.
5. The equipotential lines become more and more similar to circles as they get nearer to either charge.
6. Just behind the charges the effect of the opposite charge is minimal so the field lines resemble those of an isolated charge.

The better you grasp the salient points (and store them in your memory), the more physical intuition and predictive power you will acquire. Shouldn't one rely solely on mathematics? Physical intuition might provide *part* of the answer, but surely mathematics will always give the *complete* answer. This is true for simple problems but, as I have said many times before, as soon as the

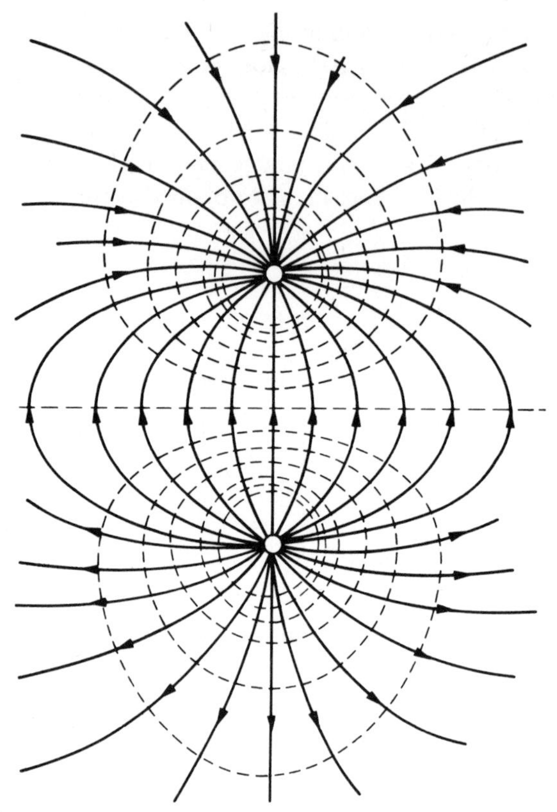

FIG. 2.4. The field lines and equipotentials for two equal charges of opposite sign.

problems become more complicated our mathematical knowledge turns out to be greatly deficient.

In practice you will find that problems may be broadly divided into two classes: (I) problems that have been solved, and (II) problems that are insoluble. The reasons for belonging into class II may be numerous: you cannot formulate the problem, you cannot solve the resulting equations analytically, the computer you have access to is not big enough, etc., etc. It all boils down to the fact that the problem needs to be simplified. You will find that you never solve the original problem. At best you solve a similar one. But how can you recognize a 'similar' problem? Which simplifications are admissible without losing the essential characteristics of the problem? For all that you need intuition.

The electric dipole

One often goes to extremes in order to arrive at a physical configuration that is mathematically soluble by simple means. This is what we are going to do now and besides assuming that $q = q_1 = -q_2$ we will also state that the two opposite charges are infinitesimally close to each other. Is such a situation entirely fictitious? No, it can occur in practice. But surely, two charges will never be infinitesimally close to each other. Of course not, all I mean is that the distance between the charges is very small with respect to some other distance of interest. For example, the two charges may be of some atomic distance (of the order of 10^{-10} m) apart, whereas we are interested in their effect at macroscopic distances; or take the so-called dipole aerial where the assumed separation of charges is small in comparison with the wavelength of oscillation. We shall return to the latter, time-varying problem in Section 5.16; let us first solve here the static case.

If we are considering distances far away from the origin of Fig. 2.3, i.e.

$$r = (x^2 + y^2 + z^2)^{\frac{1}{2}} \gg d, \tag{2.27}$$

using the approximation

$$(1+\delta)^n \cong 1 + n\delta, \qquad \delta \ll 1 \tag{2.28}$$

we obtain

$$\left[x^2 + \left(y - \frac{d}{2}\right)^2 + z^2\right]^{-\frac{1}{2}} - \left[x^2 + \left(y + \frac{d}{2}\right)^2 + z^2\right]^{-\frac{1}{2}} \cong \frac{yd}{r^3} \tag{2.29}$$

with which eqn (2.26) reduces to

$$\phi = \frac{y}{4\pi\varepsilon_0 r^3} qd. \tag{2.30}$$

The product qd is called the *electric dipole moment*, and we shall denote it by p_e. We shall also introduce a vector $\mathbf{p}_e = q\mathbf{d}$, where $\mathbf{d}$ is the vector connecting the two point charges located at $y = \pm d/2$. By definition the vector points from the negative to the positive charge. With the aid of this vector we can replace yqd/r by $\mathbf{p}_e \cdot \mathbf{i}_r$, where $\mathbf{i}_r$ is the unit vector in the direction of point P at a radius r (Fig. 2.5). So we may write eqn (2.30) in the form

$$\phi = \frac{\mathbf{p}_e \cdot \mathbf{i}_r}{4\pi\varepsilon_0 r^2}. \tag{2.31}$$

Yet another alternative form is obtained by using spherical coordinates (r, θ, φ) in which ϕ appears as

$$\phi = \frac{p_e \cos\theta}{4\pi\varepsilon_0 r^2}. \tag{2.32}$$

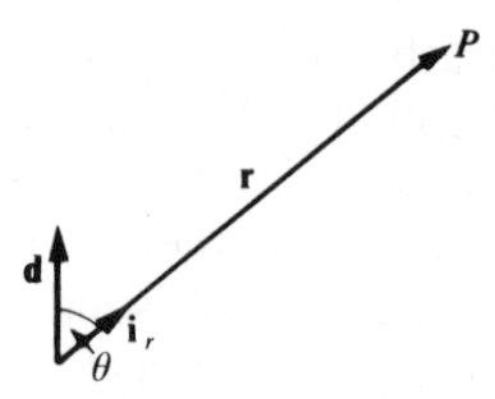

FIG. 2.5. Vectors appearing in the dipole problem.

The above form is probably best suited for deriving the components of the electric field. Using the formulae (A−11 and 14) for the gradient in spherical coordinates, we get

$$E_r = -\frac{\partial\phi}{\partial r} = \frac{p_e}{4\pi\varepsilon_0}\frac{2\cos\theta}{r^3},$$

$$E_\theta = -\frac{1}{r}\frac{\partial\phi}{\partial\theta} = \frac{p_e}{4\pi\varepsilon_0}\frac{\sin\theta}{r^3}, \tag{2.33}$$

$$E_\varphi = 0.$$

The potential far away from the charges

If the charges are within a finite volume τ and we wish to determine the potential at a point P, far away from this volume, the formula we have obtained before (eqn (2.23)) may be simplified.

We shall assume (Fig. 2.6) that the vectors drawn from any point charge to point P are all parallel to the vector $\mathbf{r}_0$ drawn from an arbitrarily chosen origin inside τ (just another way of saying that the point P is far away). This is a technique often used that leads to a quick and simple answer. We may then express the individual distances as

$$|\mathbf{r}_i| \cong |\mathbf{r}_0| - \mathbf{d}_i \cdot \mathbf{i}_{r0}, \tag{2.34}$$

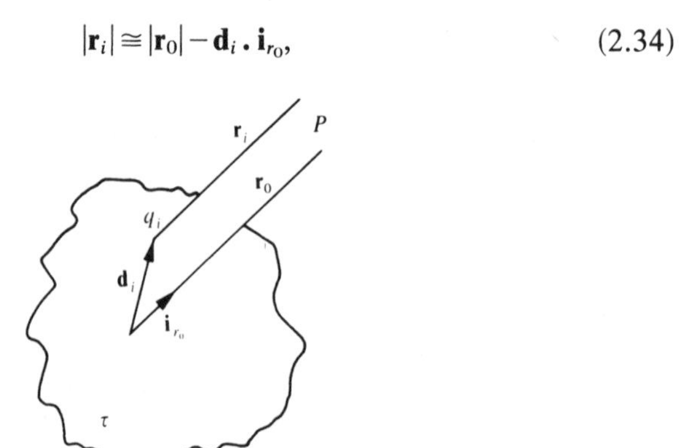

FIG. 2.6. All charges contained by the volume τ; potential to be determined at a distance far away from the charges.

where $\mathbf{d}_i$ is the vector giving the position of charge q_i. With the aid of the above relation we get

$$\frac{1}{r_i}=\frac{1}{r_0-\mathbf{d}_i\,.\,\mathbf{i}_{r0}}\cong\frac{1}{r_0}\left(1+\frac{\mathbf{d}_i\,.\,\mathbf{i}_{r0}}{r_0}\right), \tag{2.35}$$

which substituted into eqn (2.23) yields

$$\phi=\frac{1}{4\pi\varepsilon_0}\left\{\frac{1}{r_0}\sum_{i=1}^{n}q_i+\frac{1}{r_0^2}\sum_{i=1}^{n}q_i\mathbf{d}_i\,.\,\mathbf{i}_{r0}\right\}. \tag{2.36}$$

Thus we have a contribution depending on the total charge, and a second term bearing strong resemblance to our dipole formulae. Since the second term decays as $1/r_0^2$, could it ever become important? Yes, for many charge distributions of practical interest the net charge is zero, so the first term disappears and the second term acquires significance.

For two charges of opposite sign eqn (2.36) reduces to eqn (2.31); we just have the dipole potential previously derived. For a large number of charges we sum up the contribution of each dipole moment.

Multipoles

Why stop at dipoles? Could we have higher-order moments as well? The answer is yes, but the mathematics gets more and more tedious. A not-too-difficult example is a special sort of quadrupole (two dipoles of equal dipole moment and of opposite directions arranged axially) shown in Fig. 2.7. Then

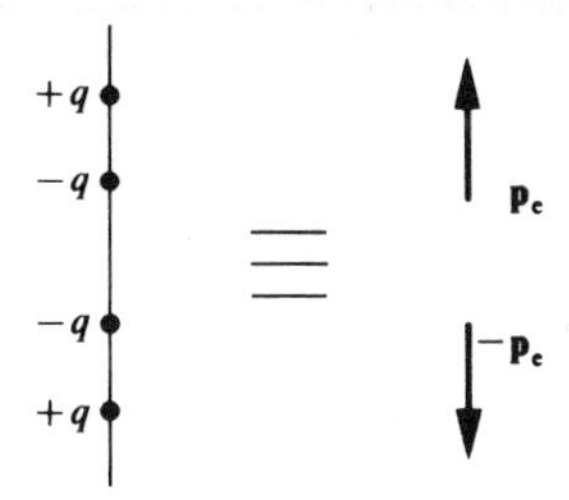

FIG. 2.7. An axial quadrupole.

eqn (2.36) would give zero and we need better approximations for r_i. If you are good at expanding functions up to second order you might like to attempt Example 2.5.

2.4. The electric field due to a line charge

Let us take as an example an infinitely long, infinitely thin distribution of charges as shown in Fig. 2.8. Let us further assume that the charge distribution is uniform and denote the charge per unit length by ρ_l (we are defining thereby a linear charge density of dimension coulomb per metre).

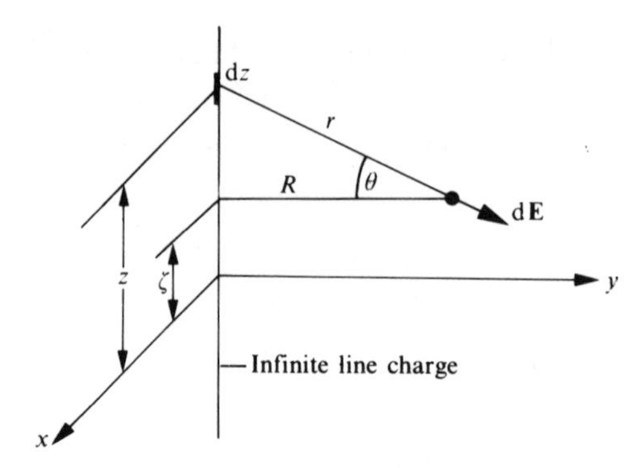

FIG. 2.8. The electric field at R, ζ due to an elementary line charge at z.

Our aim is to determine the electric field. One possible method is to work out the electric field due to a point charge $\rho_l\,\mathrm{d}z$ located at z and then add the field due to all the other point charges present. We obtain then for the electric field at the coordinate R, ζ

$$\mathrm{d}\mathbf{E} = \frac{\rho_l\,\mathrm{d}z}{4\pi\varepsilon_0 r^2}\mathbf{i}_r = \frac{\rho_l\,\mathrm{d}z\,\mathbf{i}_r}{4\pi\varepsilon_0\{(z-\zeta)^2+R^2\}}, \tag{2.37}$$

where $\mathbf{i}_r$ is the unit vector in the direction of the $\mathbf{r}$ vector. Owing to symmetry considerations (there is the same infinite amount of charge in the region $-\infty<z<\zeta$ as in the corresponding region $\zeta<z<\infty$) the electric field can have no component in the z direction, so we need only to worry about the radial component. This we may obtain easily from Fig. 2.8 as follows

$$\mathrm{d}E_R = \frac{\rho_l \cos\theta\,\mathrm{d}z}{4\pi\varepsilon_0\{(z-\zeta)^2+R^2\}}. \tag{2.38}$$

It is preferable to do the integration for θ so we shall rewrite eqn (2.38) with the aid of the relation

$$z-\zeta = R\tan\theta \tag{2.39}$$

in the form

$$\mathrm{d}E_R = \frac{\rho_l \cos\theta\,\mathrm{d}\theta}{4\pi\varepsilon_0 R}, \tag{2.40}$$

which may be integrated between the limits $\theta=-\pi/2$ and $\theta=\pi/2$ to yield

$$E_R = \frac{\rho_l}{2\pi\varepsilon_0 R}. \tag{2.41}$$

This formula tells us a lot and in simple language too. The only surviving component of the electric field varies inversely with the distance from the line charge and it is independent of the coordinate ζ. Remember, for a point charge the electric field varies with the inverse square of the distance, but it is just inverse distance for the line charge. Why is this worth remembering? Because the relationships are simple, they will not considerably burden your memory, and at the same time will assist you in building up your intuitive picture.

Take now a finite line charge as shown in Fig. 2.9(a) and (b) using different scales. Can you answer the question: what is the relative strength of the

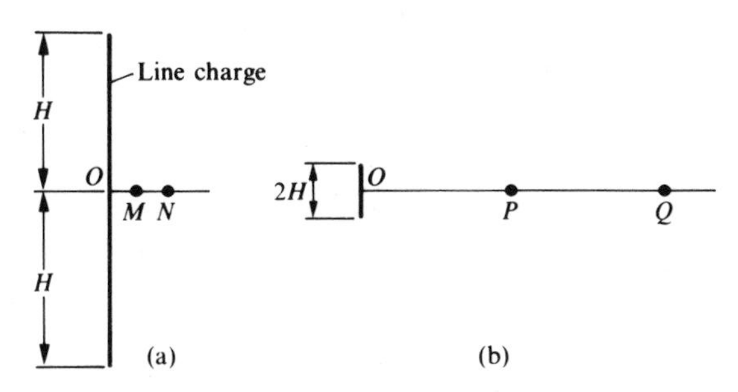

FIG. 2.9. Line charges of finite length.

electric field at points N and M in Fig. 2.9(a) and at points Q and P in Fig. 2.9(b)? To be concrete let us take

$$MO = \frac{H}{20}, \qquad NO = \frac{H}{10}, \qquad PO = 10H, \qquad QO = 20H. \qquad (2.42)$$

It looks from points N and M as if the linear charge distribution would be very long indeed, perhaps infinitely long. Thus the electric field distribution may be expected to follow a (distance)$^{-1}$ law. Hence $E_N/E_M \simeq \frac{1}{2}$. Looking from points Q and P the linear charge distribution between $z = -H$ and $z = H$ appears more like a point charge concentrated at O. Thus the electric field distribution may be expected to follow a (distance)$^{-2}$ law. Hence $E_Q/E_P \simeq \frac{1}{4}$. We have now been able to give immediate answers to fairly complicated questions. How good are the approximations? Derive the exact formula for the electric field, work out its values at points M, N, P, Q, take the ratios, and see for yourself how good our approximations are.

Let us solve now the same problem by another method. Instead of calculating the electric field directly from the charge distribution let us first determine the potential and obtain the electric field by differentiation. This is surely the better method because we won't need to worry about vectors when doing the integration. In order to avoid a lot of painful algebra we shall further simplify the problem and work out the potential for $\zeta = 0$.

The potential of a point charge is given by eqn (2.22), and that of a charge distribution by eqn (2.24). Whichever we use we end up with the following integration for the potential of an infinite line charge:

$$\phi = \frac{\rho_l}{4\pi\varepsilon_0}\int_{-\infty}^{\infty}\frac{\mathrm{d}z}{\sqrt{(z^2+R^2)}} = \frac{\rho_l}{4\pi\varepsilon_0}\left[\sin h^{-1}\frac{z}{R}\right]_{-\infty}^{\infty}. \tag{2.43}$$

All we need to do now is to put in the limits, but alas $\sinh^{-1} z/R \to \infty$ as $z \to \infty$. We get the result that the potential at point P is infinitely large. Have we done something wrong? No, we have done the same thing as before with the only difference that we calculated the potential instead of the electric field. Where could the trouble lie? Surely, in the infinite nature of our line source. There are no infinitely long line charges in nature. We have taken an unphysical picture and we get a nonsensical answer. But why did we get a reasonable result for the electric field? That was calculated for an infinitely long line charge too. Well, sometimes you get away with it, sometimes you don't. Why? The cause is only known to mathematicians and philosophers constantly engaged in the study of infinity. What can an ordinary physicist or engineer do? Well, there are several avenues open. Number one is to acknowledge the fact that our line charge is *not* infinitely long, integrate between the limits $-H$ and $+H$, differentiate to obtain the electric field, and let then H go to infinity. Then if there is any justice on earth, we shall arrive at eqn (2.41). The other thing we can do is to retrace our steps leading to eqn (2.22). We need to go only as far as eqn (2.21). We may see then that we chose our reference point at infinity. Perhaps there is the rub. We may have tampered too much with infinity. Let us abandon that assumption and see what happens. Thus we are going to say that $\phi(a) = 0$ at some other point. It probably matters little where we choose that point as long as it is not at infinity. So let us choose it for convenience at $z = 0$ at a distance R_0 from the line charge as shown in Fig. 2.10. Then the potential at P due to a point charge at z is

$$\mathrm{d}\phi = \frac{\rho_l\,\mathrm{d}z}{4\pi\varepsilon_0}\left(\frac{1}{\sqrt{(z^2+R^2)}} - \frac{1}{\sqrt{(z^2+R_0^2)}}\right), \tag{2.44}$$

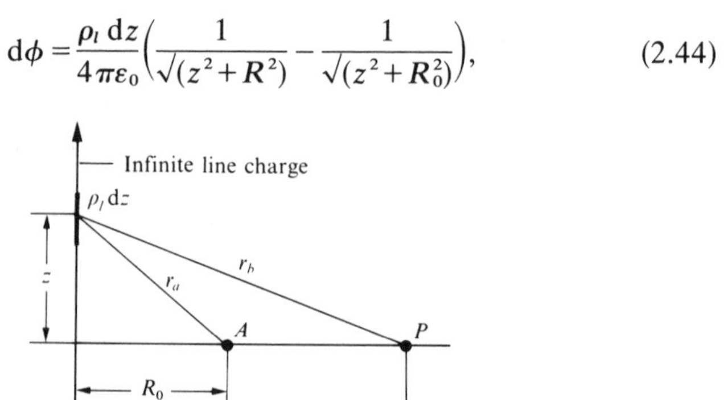

FIG. 2.10. Calculation of the potential at P; zero potential chosen at A.

and the potential due to the infinite line charge

$$\phi = \lim_{H\to\infty} \frac{\rho_l}{4\pi\varepsilon_0}\int_{-H}^{H}\left(\frac{1}{\sqrt{(z^2+R^2)}}-\frac{1}{\sqrt{(z^2+R_0^2)}}\right)\mathrm{d}z$$

$$= \lim_{H\to\infty}\frac{\rho_l}{2\pi\varepsilon_0}\left[\sinh^{-1}\frac{z}{R}-\sinh^{-1}\frac{z}{R_0}\right]_0^H$$

$$= \lim_{H\to\infty}\frac{\rho_l}{2\pi\varepsilon_0}\left[\ln\frac{\sqrt{\{1+(z/R)^2\}}}{\sqrt{\{1+(z/R_0)^2\}}}\right]_0^H$$

$$= \lim_{H\to\infty}\frac{\rho_l}{2\pi\varepsilon_0}\ln\frac{(1/R)+\sqrt{(1/H^2+1/R^2)}}{(1/R_0)+\sqrt{(1/H^2+1/R_0^2)}}$$

$$= \frac{\rho_l}{2\pi\varepsilon_0}\ln\frac{R_0}{R}. \tag{2.45}$$

Our attempts at determining both the electric field strength and the potential have started with considering the effect of a single point charge and have been followed by an integration for obtaining the total effect of the infinitely long line charge. Is there another, more direct way of determining the electric field? Well, there is Gauss's law, we haven't tried using that. If the line charge is infinitely long (so that the field depends on the radius only) we can choose our Gaussian surface as a cylinder wrapped round the line charge (Fig. 2.11). By relying on circular symmetry we can further claim that the electric field will have only a radial component which means that

$$\mathbf{E}\,.\,\mathrm{d}\mathbf{S} = E_R\,\mathrm{d}S = E_R R\,\mathrm{d}\varphi\,\mathrm{d}z, \tag{2.46}$$

where φ is the azimuth angle in the cylindrical coordinate system. The application of Gauss's law (eqn (2.15)) leads then to

$$\varepsilon_0 E_R R\int \mathrm{d}\varphi\,\mathrm{d}z = \int \rho_l\,\mathrm{d}z. \tag{2.47}$$

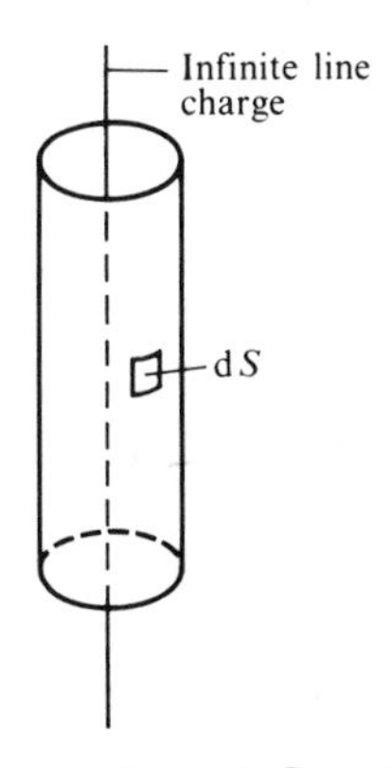

FIG. 2.11. Geometry for apply Gauss's law.

Performing the integration in the z direction for any finite length, we get for the electric field strength

$$E_R = \frac{\rho_l}{2\pi\varepsilon_0 R}, \tag{2.48}$$

in agreement with eqn (2.41).

The potential may be obtained from the electric field as

$$\phi = -\int_{R_0}^{R} \mathbf{E} \, . \, d\mathbf{S} = -\frac{\rho_l}{2\pi\varepsilon_0}\int_{R_0}^{R} \frac{dR}{R} = \frac{\rho_l}{2\pi\varepsilon_0} \ln \frac{R_0}{R}. \tag{2.49}$$

It appears that the application of Gauss's law leads much more quickly to the required result.

What is the moral of the story? (i) Tamper with infinity at your own peril, and (ii) Some ways of solving a problem are easier than others.

2.5. The electric field due to a sheet of charge

We shall now investigate the case when the charge is uniformly distributed over a plane (the x, y plane in Fig. 2.12). The charge extends to an

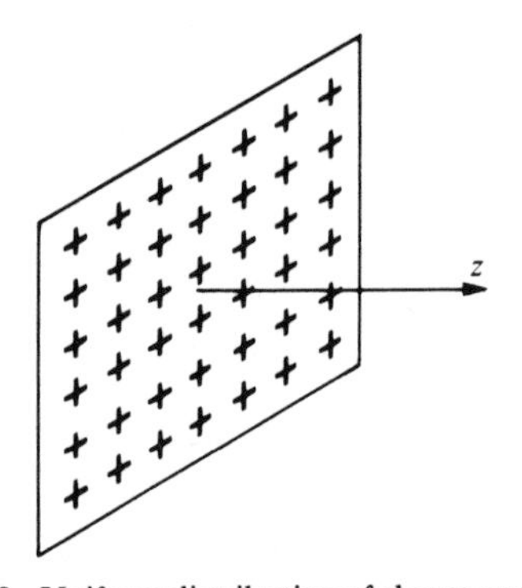

FIG. 2.12. Uniform distribution of charge over a plane.

infinitesimal distance in the z direction so we shall call it a *surface* charge. To be consistent with our previous notation for a line charge we shall denote it by ρ_s (dimension coulomb m^{-2}).†

The only component of the electric field is in the z direction (the others must be zero by symmetry considerations). If the sheet consists of positive charge the electric field points outwards. Since the space to the right of the charged sheet looks the same as that to the left, the magnitude of the electric field will be the same at $z = a$ and $z = -a$.

Learning from our meandering in the previous section we shall start straight away with Gauss's law. Choosing our surface as the box of Fig. 2.13

† The usual notation is σ but we shall reserve that symbol for the conductivity.

we get

$$\int_{S_+} \mathbf{D} \cdot \mathrm{d}\mathbf{S} = \varepsilon_0 E_z l^2, \tag{2.50}$$

$$\int_{S_-} \mathbf{D} \cdot \mathrm{d}\mathbf{S} = \varepsilon_0 E_z l^2. \tag{2.51}$$

There is no contribution from the side surfaces because the scalar product $\mathbf{E} \cdot \mathrm{d}\mathbf{S}$ vanishes.

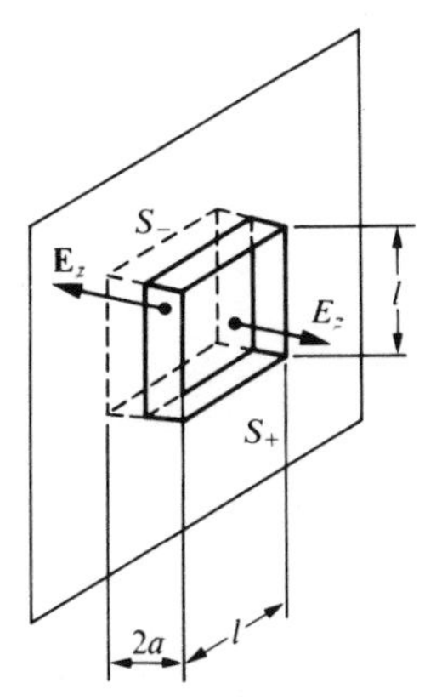

FIG. 2.13. Gaussian surface for determining the electric field.

The total charge enclosed is

$$q = \rho_s l^2, \tag{2.52}$$

hence

$$E_z = \frac{\rho_s}{2\varepsilon_0}. \tag{2.53}$$

The electric field is a constant everywhere in space but is of different sign on the two sides of the sheet.

Let us put now a sheet of opposite charge a distance d away from the first sheet (Fig. 2.14). Then the electric fields due to the two sheets add in between the sheets

$$E_z = \frac{\rho_s}{2\varepsilon_0} - \frac{-\rho_s}{2\varepsilon_0} = \frac{\rho_s}{\varepsilon_0} \quad \text{for } 0 < z < d, \tag{2.54}$$

and cancel outside the sheets

$$E_z = 0 \quad \text{for } z < 0 \text{ and } z > d. \tag{2.55}$$

The equipotential surfaces are obviously planes. We get, by integrating eqn (2.54),

$$\phi = -\frac{\rho_s}{\varepsilon_0} z + C. \tag{2.56}$$

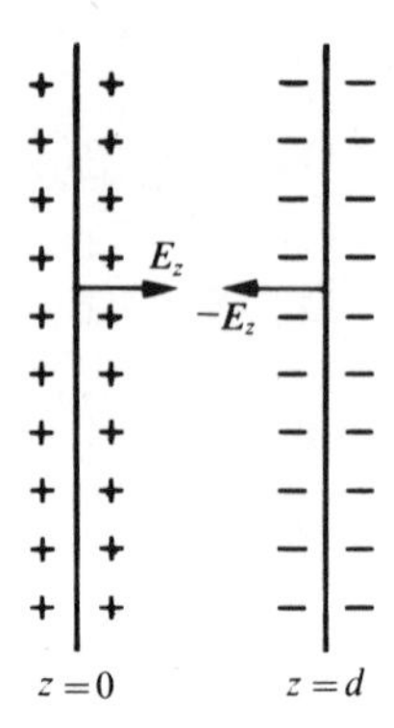

FIG. 2.14. Two sheets of charge at a distance d from each other.

2.6. The parallel-plate capacitor

We have so far considered point charges, dipoles, line charges, and sheet charges without enquiring into the problem how such configurations of charges come about. Indeed if one gives a little thought to the matter it becomes distinctly doubtful that one could ever establish anything even vaguely resembling a sheet of charge. Like charges repel each other so the charges constituting the sheet will fly apart. Interestingly, there is a way of realizing sheets of charges. We just need to apply a voltage between two metal plates.

What happens inside metal plates in response to an applied potential or for that matter what happens to charges inside any material?

There is no escape; I have to say a few words about the properties of materials. A piece of material contains equal number of positive and negative charges. If this was not so there would be large forces between various pieces of materials. To give you a feeling how large these forces could be I quote from Feynman's lectures:†

'If you were standing at arm's length from someone and each of you had *one per cent* more electrons than protons, the repelling force would be incredible. How great? Enough to lift the Empire State Building? No! To lift Mount Everest? No! The repulsion would be enough to lift a 'weight' equal to that of the entire earth!'

So as I said before a piece of material contains equal numbers of positive and negative charges. For a class of materials called conductors the internal charge distribution may be looked upon as a cloud of mobile negative particles in the background of immobile positive lattice ions. When the voltage is applied, there is initially an electric field inside the conductor and the charges start to move under its influence. So to begin with this is not a static problem at all. But if we wait patiently for a few picoseconds until the

† *The Feynman lectures on physics* (1964). Vol. II, p. 1.1. Addison-Wesley, New York.

charges rearrange themselves there will be no further motion and the problem belongs to the realm of electrostatics. But where will the electrons find their equilibrium positions? How far will they move under the influence of an attractive electric field? Disregarding the cases when the electric field is very large or the conductor is very hot (beyond the scope of this course) the electrons cannot get farther than the boundary of the conductor. So there will be an accumulation of electrons on the surface of one of the plates and consequently a deficiency of electrons at the other plate's surface. *Inside* the conductor there will be no imbalance of charge. This is an important thing to remember. If we apply a constant voltage then, after the elapse of a short time, charge neutrality is re-established in the interior of the conductor but there will be some uncompensated charges in the immediate vicinity of the surfaces. How close to the surfaces? It doesn't really matter. The scales are certainly atomic, so we are entitled to regard these charges at the surfaces as having spread out in two dimensions only. So the introduction of a surface charge is not unrealistic at all.

Well then, two infinitely large metal plates to which a constant voltage is applied are equivalent to two oppositely charged sheets. The electric field between the plates is then

$$E = \rho_s/\varepsilon_0, \tag{2.57}$$

and the potential

$$\phi = -\frac{\rho_s}{\varepsilon_0} z + \text{constant}. \tag{2.58}$$

Consequently the potential difference (i.e. the voltage) between the plates comes to

$$V = \phi(0) - \phi(d) = \frac{\rho_s}{\varepsilon_0} d = Ed. \tag{2.59}$$

Next let us work out the capacitance per unit surface area. Recall the definition from circuit theory: the capacitance is the proportionality factor relating the charge stored on one of the plates to the applied voltage, i.e.

$$q = CV. \tag{2.60}$$

The total charge per unit surface area is ρ_s, which leads to the following formula for the capacitance per unit surface area:

$$C = \frac{q}{V} = \frac{\rho_s}{\rho_s d/\varepsilon_0} = \frac{\varepsilon_0}{d}. \tag{2.61}$$

We shall keep now the voltage constant and insert a piece of dielectric between the plates as shown in Fig. 2.15. What sort of difference will that make? Since the voltage is the same and the dielectric is homogeneous the electric field is still given by

$$E = V/d. \tag{2.62}$$

The flux density, on the other hand, will be different on account of the $\mathbf{D} = \varepsilon\mathbf{E} = \varepsilon_r\varepsilon_0\mathbf{E}$ relationship. What about the surface-charge density? That will also increase by the same factor ε_r. In fact the surface-charge density will

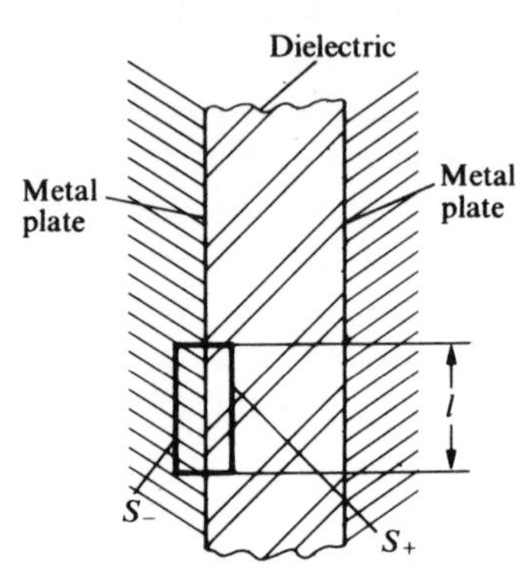

FIG. 2.15. Dielectric between two metal plates.

be equal to D. We can easily provide the proof by choosing a Gaussian surface as shown in Fig. 2.15. The contribution of S_- to the integral is now zero (because both $\mathbf{E}$ and $\mathbf{D}$ are zero inside the conductor) and that of S_+ comes to

$$\int \mathbf{D} \cdot \mathrm{d}\mathbf{S} = Dl^2. \tag{2.63}$$

Hence the application of Gauss's law yields

$$Dl^2 = \rho_s l^2 \tag{2.64}$$

or

$$\rho_s = D. \tag{2.65}$$

So we have proved that the surface-charge density increases by a factor ε_r. Why? This is a problem that rightfully belongs to the subject of the electrical properties of materials, so I cannot say very much about it.

In a perfect dielectric there are no mobile charges but lots of bound charges, positive and negative. In response to the electric field in which these charges find themselves, the positive and negative charges will slightly separate. The effect will be some uncompensated charges at the edge of the dielectric (Fig. 2.16) which will draw some further charges of the opposite sign from the interior of the conductor to its surface. This is the physical mechanism responsible for the increase of surface charge density in the presence of a dielectric.

Haven't we made a mistake? When working out the total charge within the Gaussian surface (the right-hand side of eqn (2.63)) we ignored the

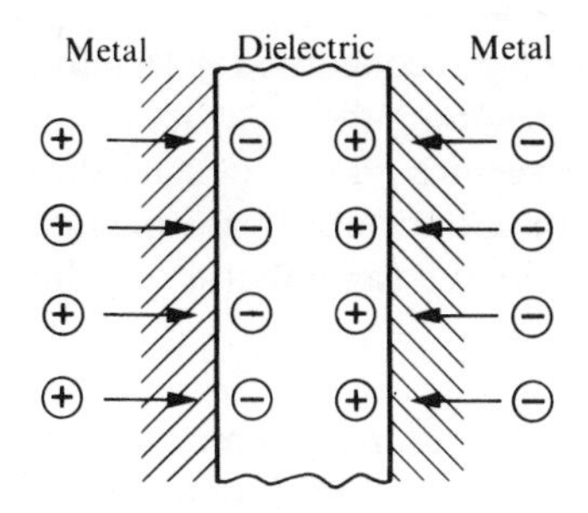

FIG. 2.16. Charges appearing on the dielectric surfaces draw further charges of opposite sign from the interior of the conductor.

charge density on the surface of the dielectric. Surely, we should have taken that into account. What counts is the *net* surface-charge density inside our chosen volume and that hasn't increased at all. All that happened is that the electric field drew some charges to the surface of the dielectric which caused then some additional charges to appear on the surface of the conductor. So the surface-charge density increased on the surface of the conductor but not the net charge density inside the Gaussian surface. Have we or have we not made a mistake? The answer depends on our definition of charge and charge density. If q in Gauss's law (eqn (2.15)) means *free* charge then we are all right, then we can ignore the *bound* charge on the dielectric. In fact this is the whole point of introducing D. By assigning a relative permittivity to a dielectric material all these problems have been taken care of.† We need to consider the free charges only.

† If we wanted to dwell more upon the fundamentals we should start with the equation

$$\varepsilon_0 \int \mathbf{E} \cdot d\mathbf{S} = q,$$

where q means any kind of charge including bound charges in a dielectric. Integrating the above equation for the Gaussian surface of Fig. 2.15 would lead to

$$\varepsilon_0 E = \rho_{s\,\text{cond}} - \rho_{s\,\text{dielec}}.$$

Thus if we identify $\varepsilon_0 E + \rho_{s\,\text{dielec}}$ with the flux density D, we get

$$D = \rho_{s\,\text{cond}},$$

which is identical with eqn (2.65). This is really the justification for using D and disregarding the surface-charge density on the dielectric. Since according to our assumptions D must be proportional to E we may write

$$\rho_{s\,\text{dielec}} = \varepsilon_0 \chi E$$

(where the proportionality factor χ is called the dielectric susceptibility), leading to

$$D = \varepsilon_0 (1 + \chi) E,$$

whence by definition

$$\epsilon_r = 1 + \chi.$$

The trouble with this attempt to go back to fundamentals is that we get only more and more entangled with the physics. Having introduced χ we could legitimately ask the question: How does χ depend on the properties of the material? If we say that χ is a constant that can be measured, we have not gained much. If we try to relate χ to some other properties of the dielectric we shall have to do some more strenuous investigations at the atomic level. As I mentioned before that is beyond the scope of these lectures, so we can just as well use ε_r as a phenomenological constant and refrain from further enquiries into the physics of dielectrics.

Let's not forget that we are concerned with parallel-plate capacitors and we are interested in determining the capacitance. Since the insertion of the dielectric increased the surface charge density on the plates by a factor ε_r with the voltage remaining unchanged, we may conclude that the capacitance has also increased by the same factor, yielding

$$C = \varepsilon_0 \varepsilon_r / d. \tag{2.66}$$

Next, we shall consider a more complicated problem where the space between the plates is filled by two different dielectrics, as shown in Fig. 2.17.

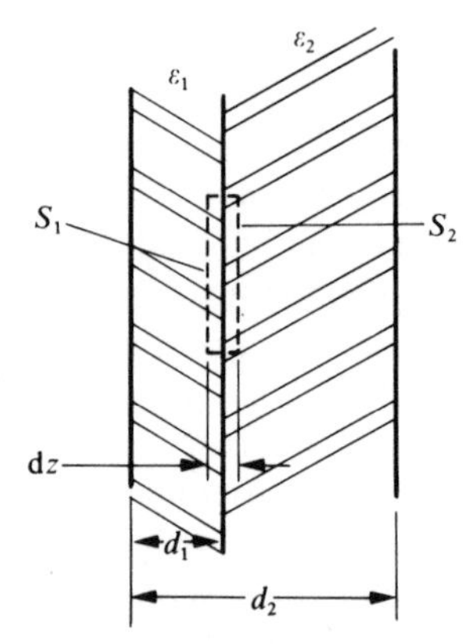

FIG. 2.17. Gaussian surface at the interface of two dielectrics.

Introducing subscripts 1 and 2 for denoting our quantities in the two dielectrics, we may write

$$D_1 = \varepsilon_1 E_1 \quad \text{and} \quad D_2 = \varepsilon_2 E_2. \tag{2.67}$$

What else do we know? In each section the relationship (2.62) must still be valid, so that

$$E_1 = \frac{V_1}{d_1} \quad \text{and} \quad E_2 = \frac{V_2}{d_2}, \tag{2.68}$$

and since potential is additive

$$V = V_1 + V_2 = E_1 d_1 + E_2 d_2 = \frac{D_1}{\varepsilon_1} d_1 + \frac{D_2}{\varepsilon_2} d_2. \tag{2.69}$$

We need one more equation between our variables that can again be provided by Gauss's law. Using the Gaussian surface shown in Fig. 2.17 and noting that

$$\mathbf{dS}_1 = -\mathbf{dS}_2, \tag{2.70}$$

we get

$$\int \mathbf{D}\,.\,\mathrm{d}\mathbf{S} = D_1 \int \mathrm{d}S_1 - D_2 \int \mathrm{d}S_2 = \text{free charge inside the Gaussian surface.} \tag{2.71}$$

The right-hand side is zero since there are no free charges inside the Gaussian surface. Hence

$$D_1 = D_2 \tag{2.72}$$

and, in view of eqn (2.65),

$$D_1 = D_2 = \rho_s. \tag{2.73}$$

With the aid of eqns (2.67) and (2.71) we may now obtain the capacitance per unit area:

$$C = \frac{q}{V} = \frac{\rho_s}{D(d_1/\varepsilon_1 + d_2/\varepsilon_2)} = \frac{1}{1/C_1 + 1/C_2}, \tag{2.74}$$

in agreement with the tenets of circuit theory.

2.7. Two-dimensional problems

We have so far had infinite metal plates and infinite dielectrics. They were chosen to be infinitely large in order to reduce the problem to a one-dimensional one. Let us take now the bold step of increasing the number of dimensions by one. What is the simplest two-dimensional problem involving conductors? Two concentric circular cylinders (Fig. 2.18) to which a voltage

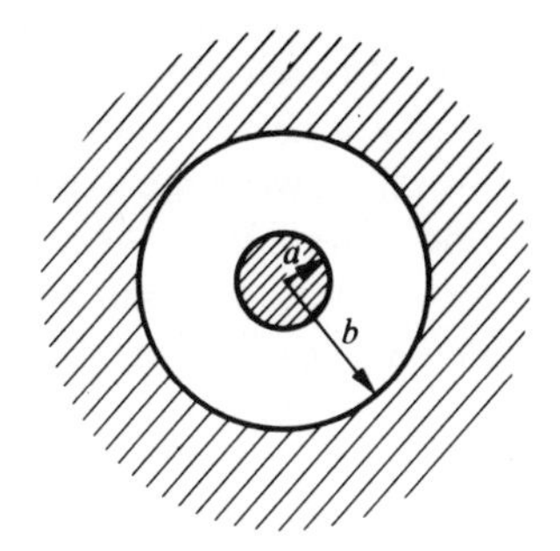

FIG. 2.18. Two concentric circular cylinders.

is applied and where the medium between the cylinders is a vacuum. Our aim is first to determine the variation of electric field as a function of radius and then to work out the capacitance per unit length.

As you are getting used to it by now we shall start with Gauss's law. The Gaussian surface will be a cylinder at radius R as shown in Fig. 2.19. Owing to circular symmetry the electric field will have only a radial component,

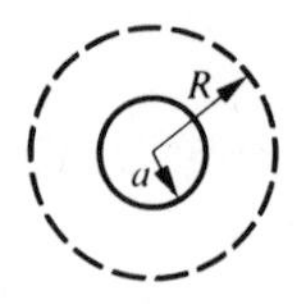

FIG. 2.19. Gaussian surface at radius R.

independent of the azimuth angle φ. Hence Gauss's law (for unit length of cylinder) takes the form

$$\int \mathbf{D} \cdot \mathrm{d}\mathbf{S} = \varepsilon_0 ER \int \mathrm{d}\varphi = 2\pi\varepsilon_0 ER = q, \tag{2.75}$$

yielding

$$E = q/2\pi\varepsilon_0 R \tag{2.76}$$

and

$$V = \int_a^b \mathbf{E} \cdot \mathrm{d}\mathbf{s} = \frac{q}{2\pi\varepsilon_0} \int_a^b \frac{\mathrm{d}R}{R} = \frac{q}{2\pi\varepsilon_0} \ln \frac{b}{a}. \tag{2.77}$$

From the definition $C = q/V$ we get immediately for the capacitance per unit length:

$$C = \frac{2\pi\varepsilon_0}{\ln (b/a)}. \tag{2.78}$$

What have we learned from the solution of our first two-dimensional problem involving conductors? We have found that the electric field varies as $1/R$. There is nothing new in that; we came to the same conclusion earlier when studying the field of a line charge. We have managed, though, to derive the capacitance of concentric cylinders, a formula used in practical engineering, so we have certainly achieved something.

It would now be easy to go on and work out the field between two concentric spheres. We shall however resist the temptation. We would learn little, because by exploiting spherical symmetry we would just have to deal with another pseudo-one-dimensional problem. It is important of course that the field varies as $1/r^2$ in that case, and it is also of some use to know the formula for the capacitance of a spherical capacitor but you can work that out yourself if you are interested.

Let us look at a *real* two-dimensional problem instead. We shall take two conducting cylinders of radius a (Fig. 2.20) and apply a voltage between them. How can we find the electric field? That should be easy. For calculating the field in point P we need two Gaussian surfaces, namely cylinders R_1 and R_2, as shown in Fig. 2.20. Then owing to the charge on one cylinder (per

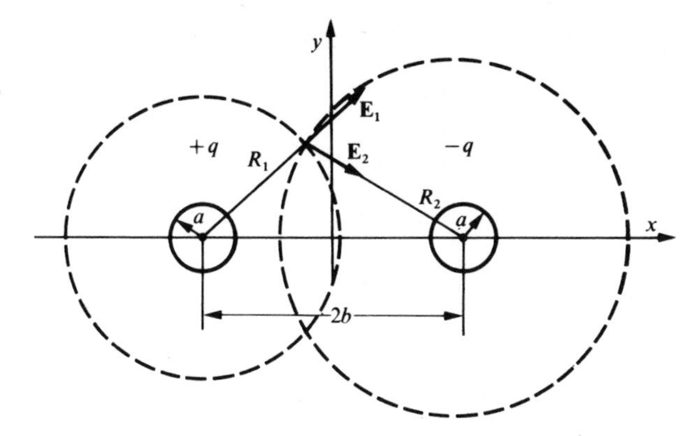

FIG. 2.20. Gaussian surfaces at radii R_1 and R_2.

unit length of course)

$$E_1 = \frac{q}{2\pi\varepsilon_0 R_1}, \tag{2.79}$$

and owing to the charge on the other cylinder

$$E_2 = \frac{q}{2\pi\varepsilon_0 R_2}. \tag{2.80}$$

The problem is linear, superposition is permissible, so all we need to do is to add vectorially $\mathbf{E}_1$ and $\mathbf{E}_2$ and the field at point P is determined.

Unfortunately the method is wrong. Why? Gauss's law is valid, and afterwards we did no more than added the field due to the two charges. This is permissible indeed if the charges are at fixed positions. In the absence of cylinder 1 the field due to the charge on cylinder 2 is correctly given by eqn (2.79). But we have used cylindrical symmetry. We have relied on the fact that the charge distribution on cylinder 1 is uniform. When we put cylinder 2 there, the circular symmetry is broken. The negative charge on cylinder 2 will attract the positive charge on cylinder 1 therefore the part of cylinder 1 facing cylinder 2 will have a higher surface-charge density than the opposite side. Gauss's law is still valid but eqns (2.79) and (2.80) do not follow from it.

How can we find a solution? Well, in this particular case we can find the solution by attacking another problem: that of two line charges as shown in Fig. 2.21. The potential for one line charge was given by eqn (2.45). For two line charges

$$\phi = \frac{\rho_l}{2\pi\varepsilon_0} \ln \frac{R_2}{R_1}. \tag{2.81}$$

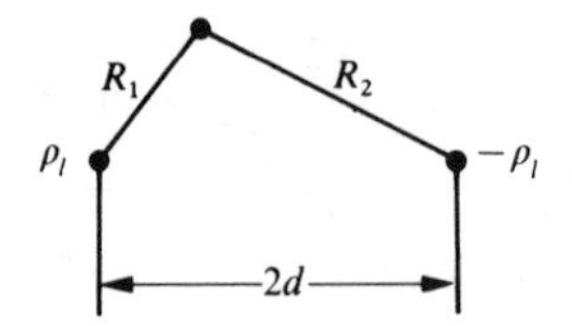

FIG. 2.21. Two line charges at a distance $2d$ from each other.

The equipotential surfaces are given by the equation

$$R_2 = \kappa R_1, \tag{2.82}$$

where κ is a constant. Doing a bit of analytical geometry it turns out that the equipotential surfaces are circular cylinders as shown by dotted lines in Fig. 2.22. We can now re-state the two-cylinder problem of Fig. 2.20 as presented in Fig. 2.23; we need only relate the parameters a, b, d, and κ to each other. The calculation outlined above yields

$$d^2 = b^2 - a^2, \tag{2.83}$$

and

$$\kappa = \frac{b+d}{a} \qquad x<0, \tag{2.84}$$

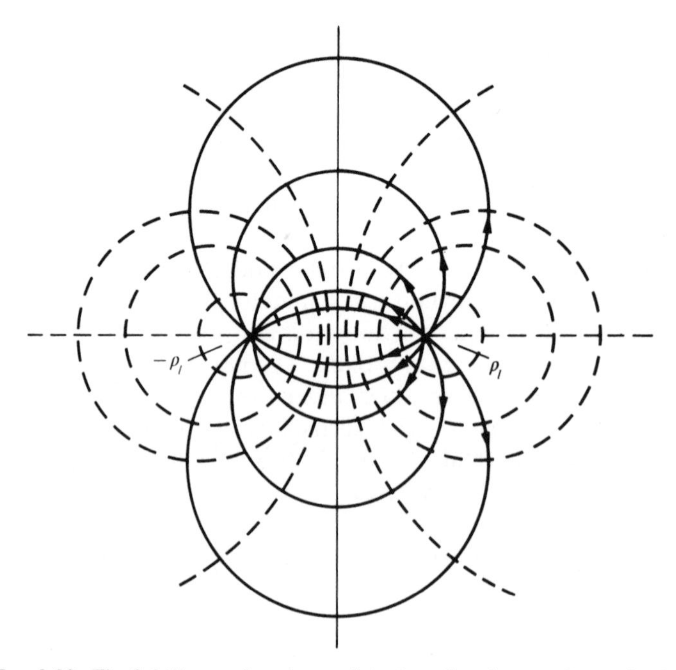

FIG. 2.22. The field lines and equipotentials of two line charges of opposite sign.

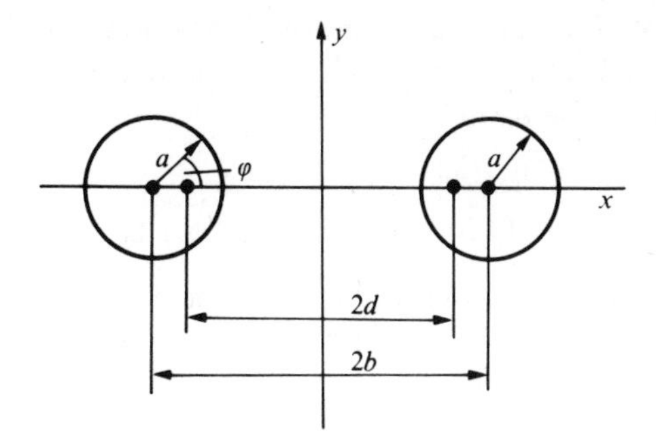

FIG. 2.23. Two charged cylinders relative to two hypothetical line charges.

$$\kappa = \frac{a}{b+d} \qquad x > 0. \tag{2.85}$$

Thus the solution for the two-cylinder problem is provided by the solution for the two-line-source problem having the same amount of charge per unit length.

We may now work out the capacitance if we wish. The potential difference may be obtained as follows

$V = \phi$ (left-hand cylinder) $- \phi$ (right-hand cylinder)

$$= \frac{\rho_l}{2\pi\varepsilon_0} \ln \kappa - \frac{\rho_l}{2\pi\varepsilon_0} \ln \frac{1}{\kappa} = \frac{\rho_l}{\pi\varepsilon_0} \ln \kappa, \tag{2.86}$$

and the capacitance per unit length

$$C = \frac{\pi\varepsilon_0}{\ln\{[b + (b^2 - a^2)^{\frac{1}{2}}]/a\}}. \tag{2.87}$$

If $b \gg a$ we get

$$C = \pi\varepsilon_0 / \ln \frac{2b}{a}. \tag{2.88}$$

Having got so far it might be of interest to work out the variation of surface-charge density on one of the cylinders as a function of φ. In order to obtain the surface-charge density we need to work out D and E on the surface of the cylinder. It is very simple in principle; we just need to use the gradient relationship between ϕ and E yielding

$$|\mathbf{E}| = \left\{ \left(\frac{\partial \phi}{\partial x} \right)^2 + \left(\frac{\partial \phi}{\partial y} \right)^2 \right\}^{\frac{1}{2}}. \tag{2.89}$$

The quantity of interest is the relative variation of the surface-charge density, i.e. $|\mathbf{D}(\varphi)|/|\mathbf{D}(0)|$, which may be obtained (after a fair amount of tedious algebra) in the form

$$\frac{|\mathbf{D}(\varphi)|}{|\mathbf{D}(0)|}=\frac{\kappa-1}{(\kappa^2-2\kappa\cos\varphi+1)^{\frac{1}{2}}}. \tag{2.90}$$

The above equation is plotted in Fig. 2.24 for $b/a=1{\cdot}1$, $1{\cdot}5$, 3, and 10. It may be seen that for a low value of b/a the surface-charge density changes considerably around the circumference of the cylinder. When $b/a \gg 1$ there is hardly any variation. So it is clear that our first approach (demonstrated in Fig. 2.20) had no general validity. But, if $b/a \gg 1$ the charges on the two cylinders have little effect upon each other, and the approach is permissible.

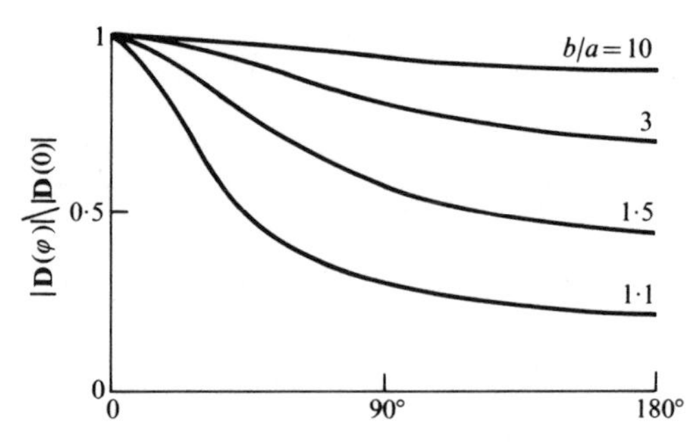

FIG. 2.24. The relative value of flux density as a function of angle round the cylinder.

We shall finish off the two-cylinder problem by working out the capacitance when $b/a \gg 1$ on the basis of eqns (2.79) and (2.80). On the line connecting the centres of the cylinders E_1 and E_2 add algebraically so that

$$E=E_1+E_2=\frac{q}{2\pi\varepsilon_0}\left(\frac{1}{b+x}+\frac{1}{b-x}\right) \tag{2.91}$$

and

$$V=\int_{-(b+a)}^{b-a} E\,\mathrm{d}x=\frac{q}{\pi\varepsilon_0}\ln\left(\frac{2b-a}{a}\right)\cong\frac{q}{\pi\varepsilon_0}\ln\frac{2b}{a}, \tag{2.92}$$

leading to a capacitance in agreement with eqn (2.88).

Having solved the two-cylinder problem we are not much wiser; we still don't know how to tackle a general problem. There is though one general rule we can set up easily enough. The electric field must always be perpendicular to the surface of a conductor. The reason is that otherwise the surface charges would be in motion contrary to the assumption that a static equilibrium exists. If the electric field is perpendicular to a surface then, owing to the $\mathbf{E} = -\nabla\phi$ relationship, that surface must be an equipotential. So if we work in terms of the potential the boundary condition may be easily formulated: ϕ = constant on all conductor surfaces. A further advantage of using ϕ (as I mentioned several times before) is that we get rid of the vectorial nature of the problem.

The mathematical problem is then to solve the differential equation

$$\nabla^2\phi = -\rho/\varepsilon \tag{2.93}$$

in conjunction with the boundary condition that ϕ = constant on all conductor surfaces.

Are there any general methods for solving eqn (2.93)? There aren't any. However, if there is no free charge in the space between the conductors, and strictly for two-dimensional problems, there is a method to which the adjective 'general' might be attached, a method that provides plenty of answers but not necessarily to the questions asked.

The method is based on the theory of complex variables. You have heard about complex variables, and I believe you have come across the Cauchy–Riemann relationships.† So you know that if

$$z = x + \mathrm{j}y \tag{2.94}$$

is a complex variable and

$$w = f(z) = u(x, y) + \mathrm{j}v(x, y) \tag{2.95}$$

is a complex function, then the relationships

$$\frac{\partial u}{\partial x} = \frac{\partial v}{\partial y} \tag{2.96}$$

and

$$\frac{\partial v}{\partial x} = -\frac{\partial u}{\partial y} \tag{2.97}$$

† Talking of Cauchy I have to tell you that he started life as an engineer. His first assignment was to help preparing the French Navy for the invasion of Britain. As so often happens in life his first job was a less than total success. His proficiency improved though later and he managed to get a number of things named after him.

I mention Cauchy to counteract the belief (prevalent in this country) that engineering is a low-prestige profession. And Cauchy is only a random example. I could quote quite a number of French engineers whose name will probably sound familiar to you, like Coulomb, Poisson, Fresnel, Carnot, Dulong, Petit, Gay-Lussac, Liouville, Becquerel, some of whom you may not immediately associate with engineering. In fact, any reasonable scientist in nineteenth-century France was likely to have had an engineering education. And this applies not only to the nineteenth century, and not only to science; out of the three serious contenders in the last French presidential election two had been graduates of the Ecole Polytechnique.

If the French tradition (that high offices should be filled with engineers) is ever going to be combined with the British tradition (that Prime Ministers should come from Oxford) then the likelihood of lecturing right now to a future Prime Minister must be considerable hold. Differentiating eqn (2.96) with respect to x and eqn (2.97) with respect to y we get

$$\frac{\partial^2 u}{\partial x^2}+\frac{\partial^2 u}{\partial y^2}=0, \tag{2.98}$$

whereas differentiation in the opposite order leads to

$$\frac{\partial^2 v}{\partial x^2}+\frac{\partial^2 v}{\partial y^2}=0. \tag{2.99}$$

Hence both u and v are solutions of Laplace's equation.

This is sheer luck to find a solution so easily. In fact we are even luckier. It turns out (though I am not going to prove it here) that the $v(x, y)=\text{constant}$ curves are orthogonal trajectories of the $u(x, y)=\text{constant}$ curves. Thus if we identify one of the functions with the potential, the other one will represent the field lines. It is really as simple as that. Take any reasonable function of a complex variable and we have the solution of an electrostatic problem.

What is the simplest function we can take? Probably

$$w=f(z)=z=x+\mathrm{j}y. \tag{2.100}$$

Take $u_1=x_1$ and $u_2=x_2$ as conductor surfaces (Fig. 2.25) then the $y=\text{constant}$ lines will represent the field lines. This example does not offer

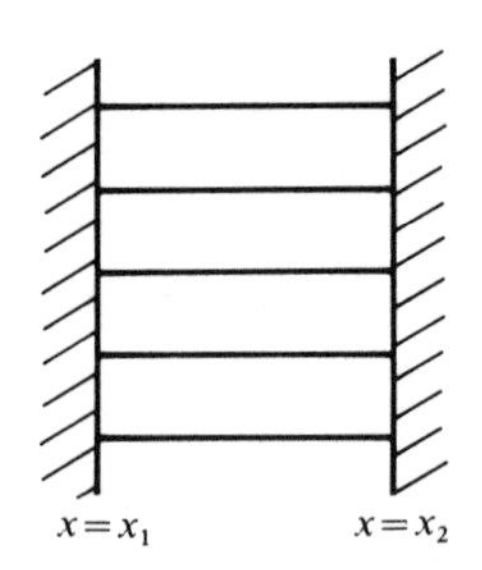

FIG. 2.25. The $y=\text{constant}$ lines.

anything new (we just got the configuration of an infinitely large parallel-plate capacitor), but we can see that the method works. Let us try as our next example something more difficult like

$$w=z^2=x^2-y^2+\mathrm{j}2xy. \tag{2.101}$$

Identifying now $v=2xy$ with the potential and $u=x^2-y^2$ with the field lines we get two sets of orthogonal hyperbolas. Taking for example $v=1$ and 2 for the conductor surfaces we have solved an electrostatic problem as may

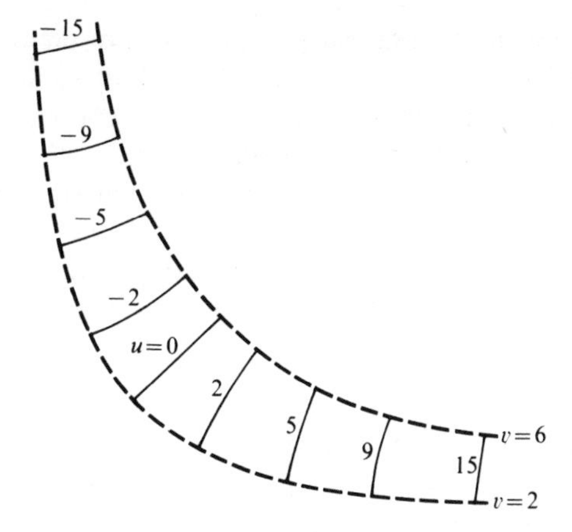

FIG. 2.26. The v = constant and u = constant lines for the complex function of eqn (2.101).

be seen in Fig. 2.26. Would anyone ever have to do anything with infinite conductors shaped like that? Very unlikely. But a finite corner is of interest, after all we often have electric fields in metal boxes. So we can take $v=0$ (giving the asymptotes) as one of the conductors and the ensuing picture (Fig. 2.27) does indeed give some intuitive 'feel' for the electric field lines in the vicinity of a corner. The electric field itself may be found from the equations

$$E_x = -\frac{\partial v}{\partial x} = -2y \quad \text{and} \quad E_y = -\frac{\partial v}{\partial y} = -2x. \qquad (2.102)$$

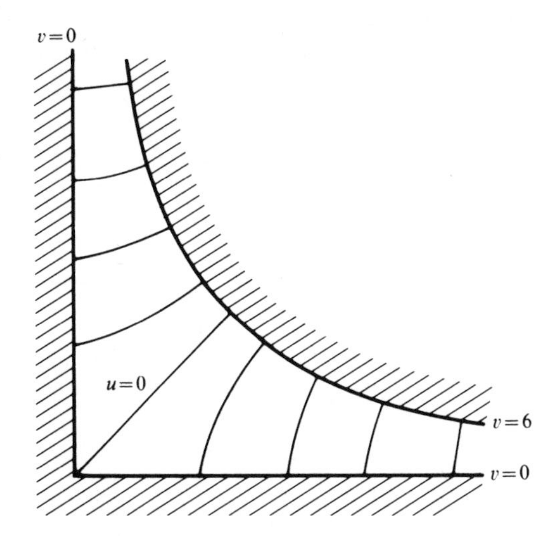

FIG. 2.27. As Fig. 2.26 but for different values of v.

If you are fond of mathematical games there are plenty of functions to experiment with. You can be sure that you will find a solution of Laplace's equation satisfying the boundary conditions, but you will have to find out which problem you have obtained the solution of.

I will not dwell much longer on *conformal mapping* (this is incidentally the term most often used for describing the method) but will give a few more examples. Let us place two coplanar conducting plates very close to each other (Fig. 2.28) and apply a potential difference between them. At some

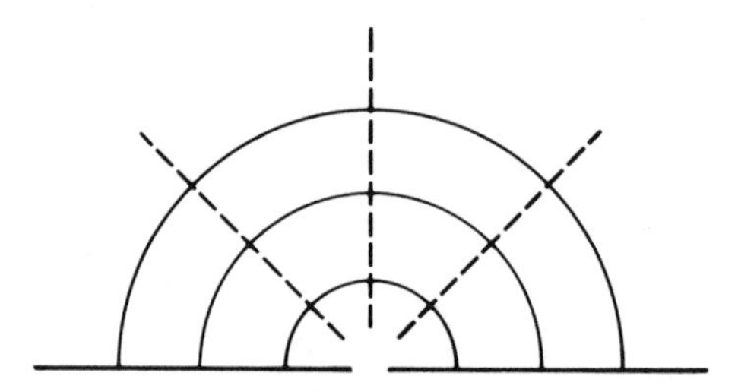

FIG. 2.28. Field lines and equipotentials for two closely placed coplanar plates at different potentials.

distance away from the gap the equipotentials will be planes and the field lines will be circles. In the immediate vicinity of the gap the situation is more complicated, as shown in Fig. 2.29. The equipotentials are confocal hyperbolas and the electric lines of force are given by confocal ellipses.

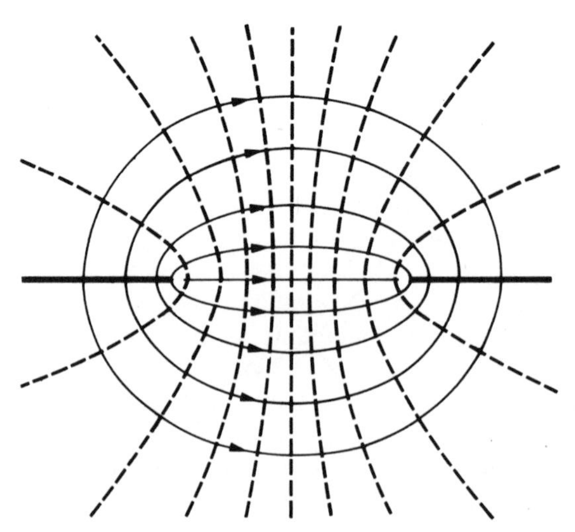

FIG. 2.29. Field lines and equipotentials in the vicinity of the gap between the coplanar plates.

Fig. 2.30 shows the field lines and equipotentials for two semi-infinite parallel conducting plates raised to different potentials. It tells us what happens at the edge of a capacitor and can also give a numerical estimate of the scattered capacitance (by which the capacitance of a real capacitor differs from that worked out on the basis of the infinite-plate model).

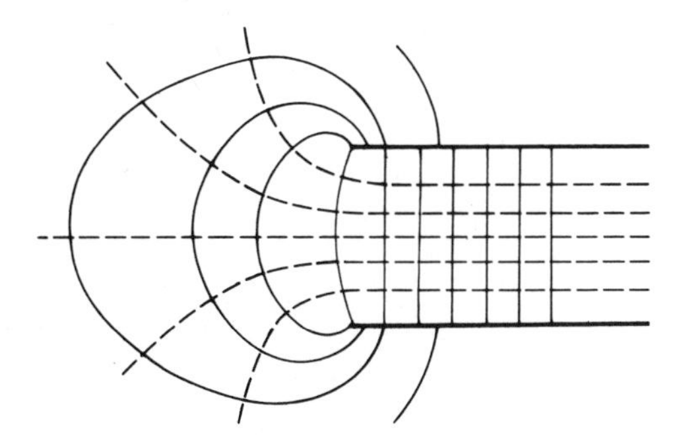

FIG. 2.30. Field lines and equipotentials at the open end of a capacitor.

The foregoing three examples provided further illustrations of the usefulness of complex functions for solving electrostatic problems (for further information on these mappings see, for example, Simonyi (1963)).† For axially symmetric cases the method is unfortunately not applicable. I shall not be discussing here the methods that are applicable but shall present the solution to one particular problem arising in electron optics, that of two closely spaced conducting cylinders at different potentials (Fig. 2.31).

2.8. The method of images

This is again a rather specific method suitable for the solution of a limited class of problems in both two and three dimensions. The simplest way of introducing the method is to present again the field lines and equipotential surfaces for two equal charges of opposite sign (Fig. 2.4). Let us now place an infinite conductor plane halfway between the charges as shown in Fig. 2.32. Since the $y=0$ plane was an equipotential surface anyway (the field lines were perpendicular to it), nothing changes. Hence we have found the solution for a charge in front of an infinite plane. Working backwards we may now say that the effect of an infinite conducting plane is equivalent to that of a charge of opposite sign placed in the mirror position (the negative charge is the *image* of the positive charge in the plane).

As an example let us work out the surface-charge density on the surface of the conducting plane due to a positive charge q above it (Fig. 2.33(a)). This is a difficult boundary-value problem, but using the method of images we only

† K. SIMONYI (1963). *Foundations of electrical engineering.* Pergamon Press, Oxford.

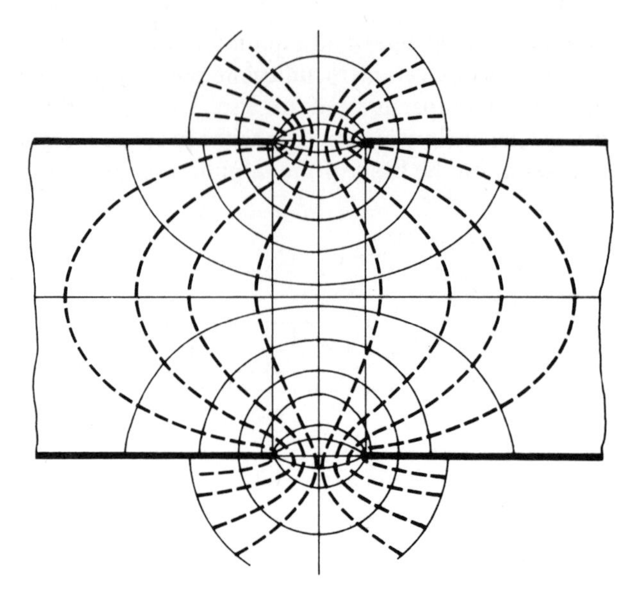

FIG. 2.31. Field lines and equipotentials due to two concentric cylinders of equal radii.

need to determine the electric flux density due to the two point charges. According to Fig. 2.33(b) the electric field at a distance r from the charge is given by

$$E = -2E_{+}\cos\theta = -2\frac{q}{4\pi\varepsilon_0 r^2}\frac{d}{r} = -\frac{qd}{2\pi\varepsilon_0 r^3} \tag{2.103}$$

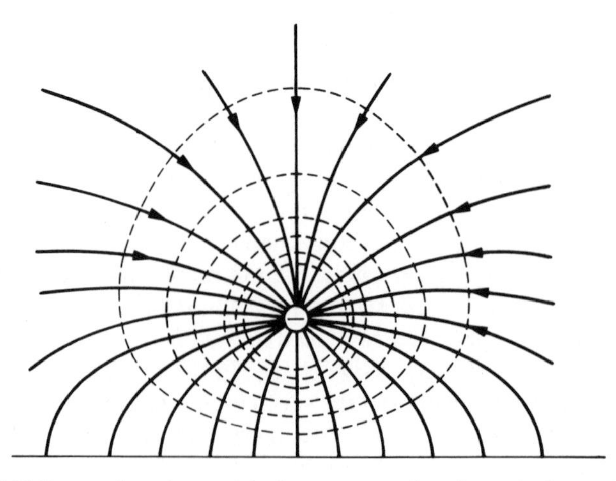

FIG. 2.32. Field lines and equipotentials due to a negative charge in front of an infinite conducting plate.

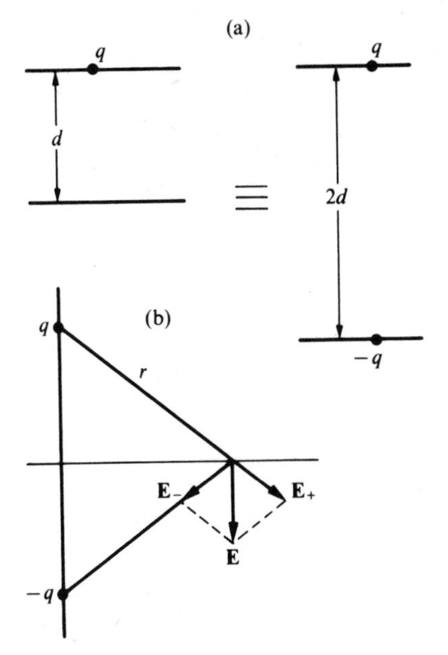

FIG. 2.33. (a) A charge at a distance d from an infinite conducting plate is equivalent to two point charges of opposite sign at a distance $2d$ from each other. (b) The electric field due to the two point charges.

(the negative sign is due to the fact that the direction of the electric field is opposite to the normal to the plane), and consequently

$$\rho_s = \varepsilon_0 E = D = -\frac{qd}{2\pi r^3}. \qquad (2.104)$$

There are various generalizations of the method. A charge in a corner has three images (Fig. 2.34) but there are as many as five images in a wedge of 60° (Fig. 2.35). You can now work out for yourself how many images a charge in a wedge of angle $2\pi/n$ has (unfortunately n must be an integer).

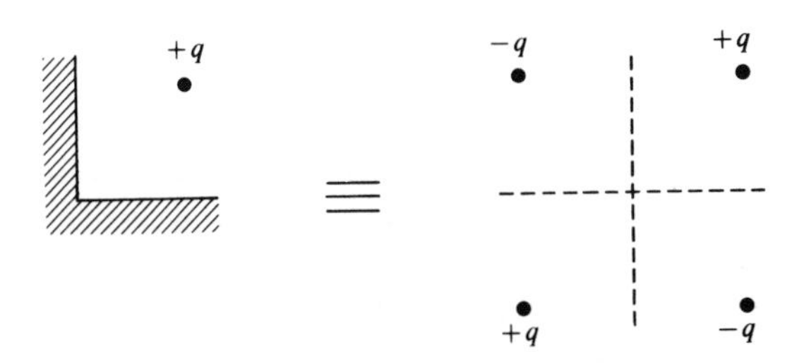

FIG. 2.34. A charge in a corner is equivalent to four charges.

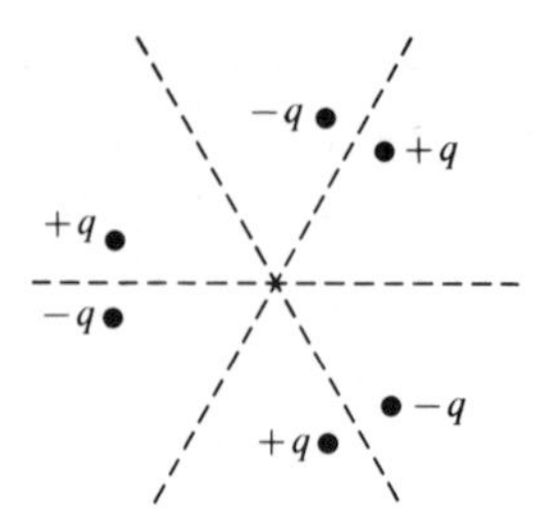

FIG. 2.35. A charge inside a wedge of 60° is equivalent to six charges.

The essence of the method is that a conductor may be replaced by a set of charges without any change in the pattern of field lines and equipotentials. Another example of a mirror charge, this time in a cylinder, is provided by Figs 2.22 and 2.23, redrawn in a more suitable form in Fig. 2.36. This is an example we have already worked out. According to our new interpretation the line charge ρ_l has its mirror charge $-\rho_l$ at a distance $2d$. It is also possible to define a mirror charge in a conducting sphere and in some purely dielectric configurations but we will not discuss them here.

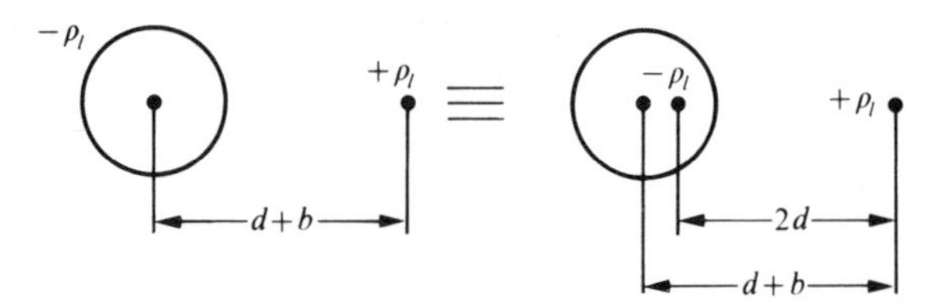

FIG. 2.36. A conducting cylinder and a line charge are equivalent to two line charges.

2.9. Dielectric boundaries

The conditions for conducting boundaries have been simple enough. Unfortunately not all boundaries are comprised of conductors; some of them are dielectrics, and even worse we have to consider sometimes imperfect dielectrics that have a finite conductivity. The proper boundary conditions may be derived in the usual manner with the aid of Gauss's law.

In the general case the boundaries are neither planes nor circles, but this fact does not need to bother us. If we investigate a small enough part of the boundary between two arbitrary media we can always regard the boundary as a plane surface. The Gaussian surface may then be chosen in the form of a cylinder, as shown in Fig. 2.37. If the height of the cylinder $\mathrm{d}h$ is approaching zero then the total flux is going through the top and bottom surfaces, i.e.

$$\int \mathbf{D}\cdot \mathrm{d}\mathbf{S} = (D_{n1} - D_{n2})\,\mathrm{d}S. \tag{2.105}$$

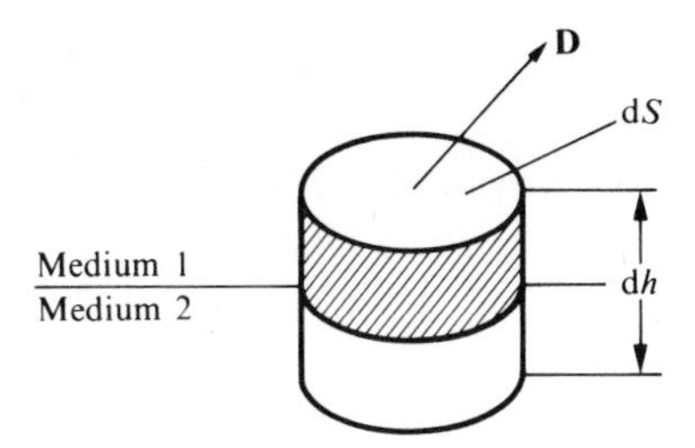

FIG. 2.37. Gaussian surface at the interface of two dielectric media.

Keeping the treatment entirely general we shall permit now the presence of a surface charge (made up of free charges; it is still true that the bound charges of dielectrics do not count), hence, when $dh \to 0$,

$$\int \rho \, d\tau = \rho_s \, dS, \tag{2.106}$$

and it follows from the above two equations that

$$D_{n1} - D_{n2} = \rho_s. \tag{2.107}$$

This is now entirely general. It includes as special cases both eqn (2.63) (when D_{n2} is zero inside the metal), and eqn (2.72) (when both dielectrics are perfect so that no free charges can reside on the boundary surface).

So far, so good. We have derived the condition for the normal component of **D**. What about the tangential component? We can get that by taking this

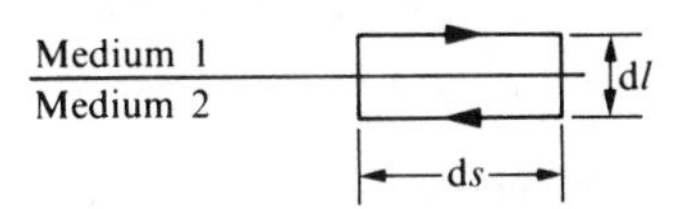

FIG. 2.38. A closed path at the interface for working out the line integral of the electric field.

time the line integral of the electric field along both sides of the boundary as shown in Fig. 2.38. Assuming that $dl \to 0$ and noting that for a closed contour the line integral of the electric field vanishes we get

$$(E_{t2} - E_{t1}) \, ds = 0, \tag{2.108}$$

i.e., the tangential component of the electric field strength is constant across a boundary.† This means that the electric field will refract at the boundary of two dielectrics, as shown in Fig. 2.39. The relevant equations are

$$E_{t1} = E_{t2} \quad \text{and} \quad \varepsilon_1 E_{n1} = \varepsilon_2 E_{n2}. \tag{2.109}$$

† For the static case this follows from eqn (2.7) when $b = a$. The integral happens to be zero for the time-varying case as well as will be discussed in Section 5.6.

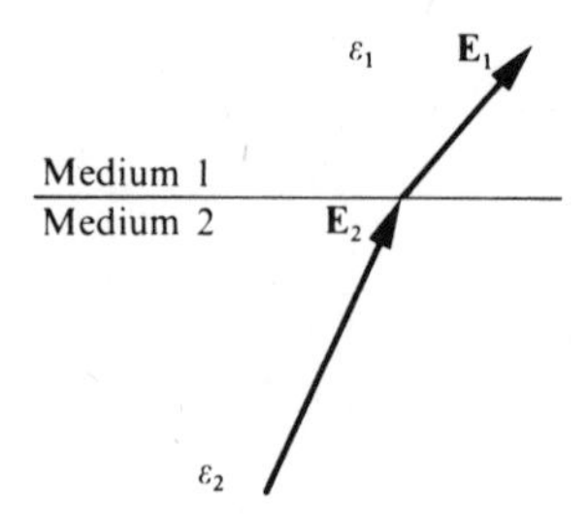

FIG. 2.39. The electric field vectors at the interface of two dielectrics.

Thus unless the incident angle is 90° the field lines will have a break at the boundary of two dielectrics.

An example is shown in Fig. 2.40. We have a line charge in front of a dielectric that fills half the space. The field lines, as expected, refract when entering the dielectric (for a mathematical solution in terms of images, see Clemmow (1973)).†

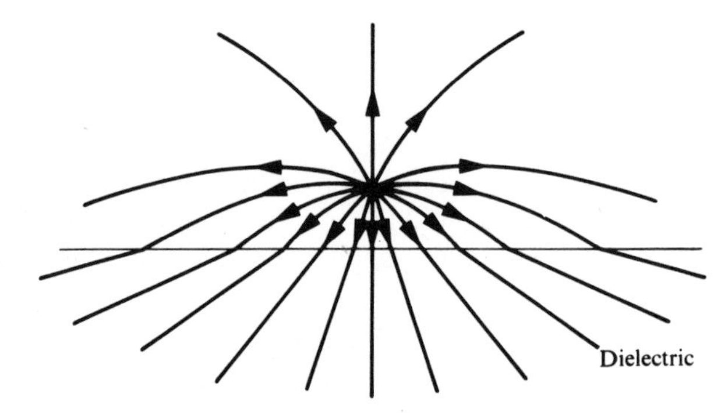

FIG. 2.40. The field lines of a point charge when half the space is filled with dielectric.

Another example demonstrating the same phenomenon may be seen in Fig. 2.41. A dielectric sphere inserted into a homogeneous electric field in air or a vacuum 'attracts' the field lines. The physical explanation is that the dielectric is capable of carrying a higher flux density. The mathematical solution is given below in spherical coordinates, r, θ (the solution is of course independent of the azimuth angle φ):

$$\phi = \begin{cases} -E_0 r\cos\theta + \dfrac{\varepsilon_r - 1}{\varepsilon_r + 2} a^3 E_0 \dfrac{\cos\theta}{r^2} & r > a, \\ -\dfrac{3}{\varepsilon_r + 2} E_0 r \cos\theta & r < a, \end{cases} \tag{2.110}$$

† P. C. CLEMMOW (1973). *An introduction to electromagnetic theory*, Section 3.4.5. Cambridge University Press.

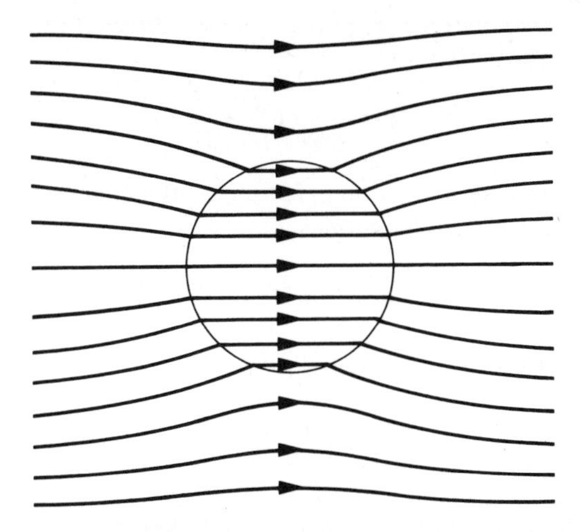

FIG. 2.41. Distortion of field lines by a dielectric sphere.

from which **E** and **D** may be derived (E_0 is the electric field strength in the absence of the sphere). The boundary conditions, eqn (2.107), are satisfied at $r = a$. If you are interested you can check them (Example 2.12).

In an isotropic medium (in which the dielectric constant is a scalar not a tensor) the directions of the **E** and **D** lines coincide, thus Figs. 2.40 and 2.41 may refer to either. However if we want to interpret the density of field lines as being proportional to the field strength then both figures must represent **D** since it is the flux of **D** that remains constant across a dielectric boundary.

2.10. Electrostatic energy

Electrostatics is characterized by charges located at certain positions. If we know the positions of all the charges, we know everything so we should know the energy as well. There is unfortunately no obvious way of writing an expression for the energy in the simultaneous presence of all the charges. It is, however, fairly easy to work out the energy as we bring in the charges from infinity one by one.

The first charge is brought in without any opposition. The energy of one charge alone is zero.† The second charge is bróught in in the presence of the first one. The work done is

$$W_2 = \frac{q_1 q_2}{4\pi\varepsilon_0 r_{12}}. \tag{2.111}$$

† The self-energy of a point charge is disregarded here. It is of no interest in engineering contexts and besides it's a problem to which no satisfactory solution exists.

There are now two charges, q_1 and q_2, and we bring a third one, q_3, from infinity. The work done is

$$W_3 = \frac{q_3}{4\pi\varepsilon_0}\left(\frac{q_1}{r_{13}} + \frac{q_2}{r_{23}}\right). \tag{2.112}$$

Three charges are probably enough for seeing the general trend so we shall now sum up the two partial energies

$$W_2 + W_3 = \frac{1}{4\pi\varepsilon_0}\left(\frac{q_1q_2}{r_{12}} + \frac{q_1q_3}{r_{13}} + \frac{q_2q_3}{r_{23}}\right), \tag{2.113}$$

which may also be written in the form

$$W_2 + W_3 = \frac{1}{8\pi\varepsilon_0}\left\{q_1\left(\frac{q_2}{r_{12}} + \frac{q_3}{r_{13}}\right) + q_2\left(\frac{q_1}{r_{12}} + \frac{q_3}{r_{23}}\right) + q_3\left(\frac{q_1}{r_{13}} + \frac{q_2}{r_{23}}\right)\right\}. \tag{2.114}$$

It may be recognized that the terms in the brackets represent the potentials due to the *other two* charges. It is easy to guess that in the general case one would have, instead, the potential due to *all* other charges. Hence the general term is of the form $q_i\phi_i$, where ϕ_i is the potential at the point where q_i resides (ignoring the contribution of q_i). The expression for the total energy of n charges is then

$$W = \frac{1}{2}\sum_{i=1}^{n} q_i\phi_i. \tag{2.115}$$

The transition from discrete to distributed charge can be made as follows: q_i is replaced by $\rho\,\mathrm{d}\tau$, ϕ_i by ϕ, and the summation by integration, yielding

$$W = \frac{1}{2}\int_\tau \rho\phi\,\mathrm{d}\tau, \tag{2.116}$$

where the volume τ must contain all the charges. Thus if the charge distribution is known we should first determine the potential (from eqn (2.24)) and then we may use eqn (2.116) for calculating the energy. This is a perfectly reasonable procedure, but we have to admit that eqn (2.116) is rarely used. It is usually transformed into another much more popular form.

Since

$$\rho = \varepsilon_0\nabla\,.\,\mathbf{E} \tag{2.117}$$

and by eqn (A.2)

$$\nabla\,.\,(\phi\mathbf{E}) = \phi\nabla\,.\,\mathbf{E} + \mathbf{E}\,.\,\nabla\phi, \tag{2.118}$$

we may first transform eqn (2.116) into

$$W = \frac{\varepsilon_0}{2}\int_\tau \{\nabla\,.\,(\phi\mathbf{E}) - \mathbf{E}\,.\,\nabla\phi\}\,\mathrm{d}\tau. \tag{2.119}$$

Noting further that $\nabla\phi = -\mathbf{E}$ and applying Gauss's theorem to the first term in the bracket we get

$$W = \frac{\varepsilon_0}{2}\left\{\int_S \phi\mathbf{E}\,.\,\mathrm{d}\mathbf{S} + \int_\tau \mathbf{E}^2\,\mathrm{d}\tau\right\}, \tag{2.120}$$

where S is the closed surface of the volume τ. Now this is another perfectly reasonable formula to use—once the volume is chosen. Well, let us choose it in the form of a big sphere. If r the radius of this sphere is big enough then the potential decays as $1/r$ (cf. Section 2.4) and the electric field as $1/r^2$. Thus the integrand decays as $1/r^3$, whereas the surface increases only as r^2. Consequently the surface integral must vanish as $r \to \infty$.

We are now left with the sought-for expression. Provided the volume integral is over all space the energy is given by

$$W = \frac{\varepsilon_0}{2}\int \mathbf{E}^2\,\mathrm{d}\tau. \tag{2.121}$$

This is a simple and physically easily interpretable expression. Wherever there is electric field, there is energy as well.

So how should you think about electrostatic energy? You are certainly entitled to think in terms of charges and potentials. In order to create a charge distribution a certain amount of work has to be done, and that is available to us in the form of electrostatic energy. But it is better (it is more general) to regard the electric field as the agent with which the energy has been deposited. One usually says that the energy is *stored* in the electric field.

Let us work out, as an example, the stored energy of a parallel-plate capacitor. The electric field strength is given (eqn (2.57)) by $E = V/d$, hence the stored energy is

$$W = \frac{\varepsilon_0}{2}\frac{V^2}{d^2}\int \mathrm{d}\tau = \frac{\varepsilon_0}{2}\frac{V^2}{d^2}S_0 d = \tfrac{1}{2}CV^2, \tag{2.122}$$

where S_0 is the area of the capacitor plates. Of course, you can derive the above expression from circuit theory, so once more circuit theory and electromagnetic theory give the same answer.† Could we use this stored energy for calculating the magnitude of forces acting, could we get the *direction* of the forces? Take the following example from mechanics: a mass m at rest at a height h has a potential energy mgh. In which direction will it move if its support is taken away? It will try to reduce the height h in order to minimize its potential energy.

† Is it something to brag about when two theories give the same answer? If one is more general than the other one (as electromagnetic theory is more general than circuit theory) then the two theories are bound to give identical answers. So objectively speaking there is no reason for rejoicing. Psychologically, on the other hand, we do occasionally need some encouragement. We live in an uncertain world, we don't know what to doubt and what to believe. Under these conditions it is positively a relief to find that two apparently independent sources come up with the same answer. Formulae derived from two different theories occupy somehow higher positions in the hierarchy of formulae. We have more faith in them.

Can we apply the same argument to the motion of capacitor plates? The stored energy, according to eqn (2.122), is $\varepsilon_0 V^2 S_0/2d$. In order to minimize this energy, d must increase. Hence, if deprived of their mechanical supports, the plates of a capacitor should fly apart. Let us check this result by invoking an alternative description of the forces in terms of surface charges. One plate is positively charged, the other plate is negatively charged; unlike charges attract each other, hence the force between the plates is attractive. Well; is the force repulsive or attractive, which argument should we believe? I would be inclined to accept the answer based on charges. Everybody knows (maybe even Arts graduates) that unlike charges attract each other, so that must be correct. What is wrong then with the energy picture? Perhaps the trouble is that while the capacitor plates move the voltage will not stay constant. But the voltage *will* stay constant; once we apply a voltage with the aid of a battery the voltage is fixed, it cannot change. So what is wrong?

Whenever you come up against a contradiction try to simplify the problem and look at it again. The obvious way of simplifying the present problem is by cutting the wires leading to the battery. In other words we propose to charge the capacitor to voltage V and then disconnect the battery. There is now no reason why the voltage should stay constant, but certainly the charge on the plates cannot change. So let us rewrite the stored energy of a capacitor in terms of the charge as follows

$$W = \frac{1}{2}\frac{q^2}{C} = \frac{1}{2}\frac{q^2 d}{\varepsilon_0 S_0}. \tag{2.123}$$

In order to minimize the stored energy the spacing between the capacitor plates has to *decrease*. The force is attractive so everything is all right.

Next we shall work out the force. Using the principle of virtual work we get

$$F = \frac{\partial W}{\partial d} = \frac{1}{2}\frac{q^2}{\varepsilon_0 S_0} = -\frac{1}{2}\frac{q^2}{C^2}\frac{\partial C}{\partial d}. \tag{2.124}$$

The contradiction has now been resolved; it is the charge that should be kept constant and not the voltage. Can we go on to the next problem? Not before making another attempt at the constant-voltage case.

When the capacitance at constant voltage changes by $\mathrm{d}C$ the energy of the capacitor increases by

$$\mathrm{d}W = \tfrac{1}{2}V^2\,\mathrm{d}C. \tag{2.125}$$

But the capacitor is *not* isolated from the rest of the universe; it is connected to a battery. So perhaps we should choose for our 'system' the capacitor and the battery together. While the energy of the capacitor increases that of the battery *decreases*. Why? Because an increase of $\mathrm{d}C$ in the capacitance requires an amount of extra charge $\mathrm{d}q = V\,\mathrm{d}C$, which must flow from the battery to the capacitor. So the *loss* of energy by the battery is $V\,\mathrm{d}q = V^2\,\mathrm{d}C$. Hence the net gain of energy of the system is

$$\mathrm{d}W = \tfrac{1}{2}V^2\,\mathrm{d}C - V^2\,\mathrm{d}C = -\tfrac{1}{2}V^2\,\mathrm{d}C = -\frac{1}{2}\frac{q^2}{C^2}\,\mathrm{d}C, \tag{2.126}$$

in agreement with eqn (2.124).

So if we consider the capacitor and the battery *together* the total energy of the system decreases as the capacitance increases. We obtain the same force between the plates whether the battery is disconnected or not—as we should.

Examples 2

1. Four point charges of equal magnitude are located at the corners of a square as shown in Fig. 2.42. Determine the magnitude and direction of the force on each charge.

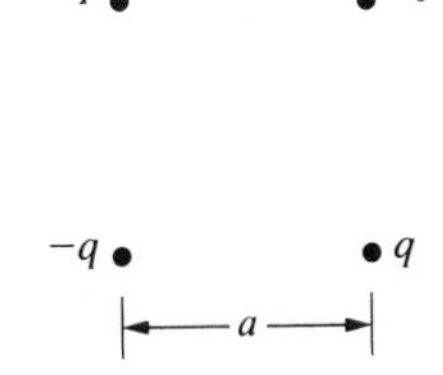

FIG. 2.42. Four point charges.

2.2. A cloud of charged particles having a total charge q, fills uniformly the volume of a sphere of radius a. Find the electric field at a distance r from the centre of the cloud both for $r < a$ and $r > a$.

2.3. Determine the electric field on the axis of a charged ring of radius a carrying a uniform line charge of ρ_l coulomb per unit length.

2.4. A linear line charge of ρ_l coulomb per unit length extends from the origin to $z = -\infty$. Determine the electric field at an arbitrary point using cylindrical coordinates R, φ, z.

2.5. Determine the electric field as a function of r and θ produced by the axial quadrupole shown in Fig. 2.43. Assume that $r \gg d$.

2.6. An infinitely long, perfectly conducting cylinder of radius a is placed into a uniform electric field perpendicularly to the direction of the field lines. The potential function for this case is given as

$$\phi = A\left(R - \frac{a^2}{R}\right)\cos\varphi,$$

where A is a constant and R and φ are polar coordinates centred at the axis of the cylinder.

(i) Show that the resultant electric field satisfies the boundary conditions,

(ii) Determine the differential equation of the field lines.

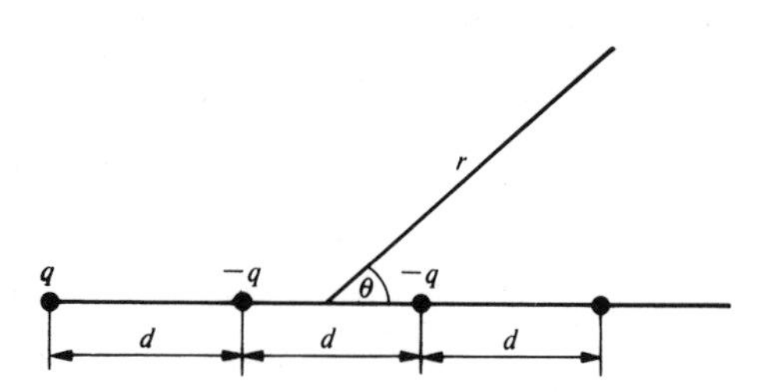

FIG. 2.43. An axial quadrupole.

2.7. Show that the potential function ϕ given in Example 6 satisfies Laplace's equation in cylindrical coordinates.

2.8. Two point charges q and $-pq$ $(p \neq 1)$ are placed at the points $(0, 0, 0)$ and $(0, 0, d)$ of a cartesian coordinate system. Find the zero potential surface.

2.9. Determine the force upon a point charge placed inside a conducting sphere at a distance a from the centre of the sphere. (*Hint*: Find the mirror charge in the sphere based on the calculations of the previous example.)

2.10. Determine the capacitance per unit length of a two-wire transmission line above a perfectly conducting earth (Fig. 2.44). Assume that d_1, $d_2 \gg D$.

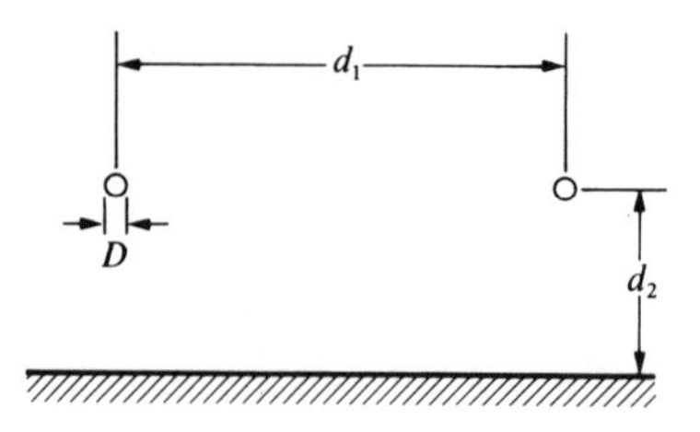

FIG. 2.44. Two parallel wires above an infinitely conducting plate.

2.11. Derive eqns (2.83)–(2.85).

2.12. Show that eqn (2.110) satisfies the boundary conditions at $r = a$.

2.13. Two long concentric cylinders of radii a and b are separated by a dielectric of relative permittivity ε_r. They are held with their common axis vertical, so as to provide a liquid-level gauge. If the capacitance of the gauge is C when a fraction α of its length is immersed, find how its sensitivity

$$\frac{\mathrm{d}C/C}{\mathrm{d}\alpha/\alpha}$$

varies with α and with the relative permittivity of the fluid.

2.14. Find the equipotentials and lines of force represented by the following conformal transformations

$$\text{(i) } w = z^{\frac{1}{2}}, \qquad \text{(ii) } w = z^{-1}, \qquad \text{(iii) } w = \ln z.$$

2.15. A capacitor made of concentric cylinders has an inner radius a, outer radius b, and length l. it is filled with a dielectric of relative permittivity ε_r.

Derive an expression for the maximum stored energy, W_{max} considering that the dielectric breaks down at a field intensity E_b. Calculate W_{max} for the case when $a = 5$ mm, $b = 10$ mm, $l = 100$ mm, $E_b = 2 \times 10^7$ V m^{-1}, $\varepsilon_r = 2{\cdot}25$.

2.16. Find the lateral force on the dielectric slab partially filling the space between two parallel capacitor plates (Fig. 2.45) having a voltage V between them.

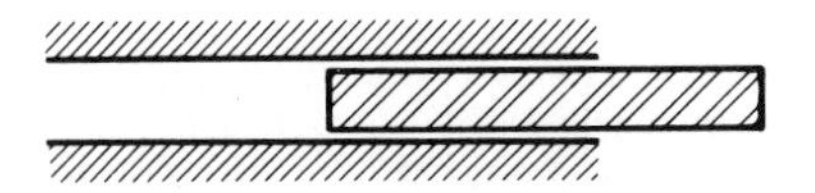

FIG. 2.45. Dielectric slab between parallel plates.

2.17. One plate of a parallel-plate capacitor having an area S is suspended at its centre from a spring of stiffness k. The other plate is held fixed at a distance d. What is the minimum voltage to be applied between the plates for pulling them together?

2.18. A variable capacitor consists of two sets of n interconnected semicircular conducting plates which can be given mutual rotation as shown in Fig. 2.46. Derive expressions for the torque needed to hold θ constant under the following conditions:

(i) when the plates are connected to a source of constant voltage V;

(ii) after the plates have been rotated to their position of maximum capacitance, charged to voltage V, disconnected from the supply and then rotated to a new position. What limits the maximum voltage that can be achieved by the operations described in (ii)?

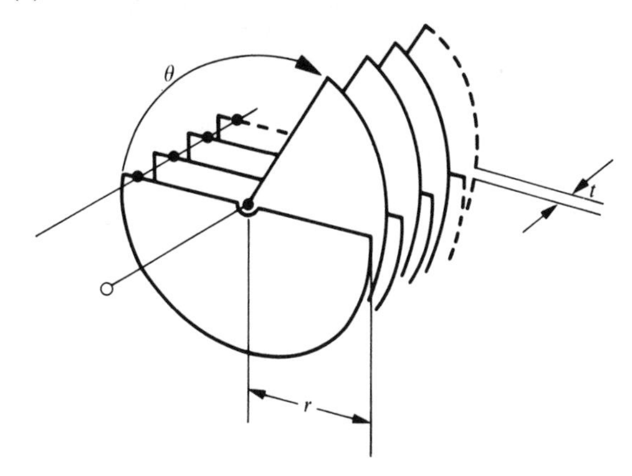

FIG. 2.46. A variable capacitor.

3. Steady currents

La force attractive & répulsive du fluide magnétique, est exactement, ainsi que dans le fluide électrique, en raison composée de la directe des densités, & inverse du carré des distances des molécules magnétiques.

CHARLES AUGUSTIN COULOMB *Second mémoire sur l'électricité et le magnetisme, Histoire de l'Académie Royale des Sciences* Paris 1785

Conjungantur termini oppositi apparatus galvanici per filum metallicum, quod brevitatis causa in posterum conductorem conjungentem vel etiam filum conjungens appellabimus. Effectui autem, qui in hoc conductore et in spatio circumjacente locum habet, conflictus electrici nomen tribuemus.

Ponatur pars rectilinea hujus fili in situ horizontali super acum magneticam rite suspensam, eique parallela. Si opus fuerit, filum conjungens ita flecti potest, ut pars eius idonea situm ad experimentum necessarium obtineat. His ita comparatis, acus magnetica movebitur, et quidem sub ea fili conjungentis parte, quæ electricitatem proxime a termino negativo apparatus galvanici accipit, occidentem versus declinabit.

Effectus fili conjungentis in acum magneticam per vitrum, per metalla, per lignum, per aquam, per resinam, per vasa figlina, per lapides transeunt; nam interjecta tabula vitrea, metallica vel lignea minime tolluntur, nec tabulis ex vitro, metallo et ligno simul interjectis evanescunt, imo vix decrescere videntur. Experimenta nostra etiam docuerunt, effectus jam memoratos non mutari, si acus magnetica pyxide ex orichalco aqua repleta includitur. Effectuum transitum per omnes has materias in electricitate et galvanismo antea nunquam observatum fuisse, monere haud opus est. Effectus igitur, qui locum habent in conflictu electrico, ab effectibus unius vel alterius vis electricæ quam maxime sunt diversi.

From a report by Hans Christian Oersted sent to a number of European Laboratories in July, 1820.

Forfatteren selv havde allerede for lang Tid siden antaget et System, i Følge hvilket alle indvortes Virkninger, i Legemerne, som Electricitet, Magnetisme, Varme, Lys, saavel som og de chemiske Forbindelser og Adskillelser beroe paa samme Grundkræfter. Dette System, som han allerede tidligere i enkelte Afhandlinger har foredraget, har han i sine *Ansichten der chemischen Naturgesetze*, der udkom 1812 fuldstændigere udviklet, og kom allerede den Gang til det Resultat, at Magnetismen maatte frembringes af de electriske Kræfter i deres meest bundne Tilstand. At bekræfte denne Tanke ved Forsøg, forestillede han sig i lang Tid vanskeligere end Udfaldet siden viste at det var. Han overloed sig derfor till andre Undersøgelser, indtil han ved sine Forelæsninger over Electricitet, Galvanismus og Magnetismus i Foraaret 1820 atter kom til at forfølge denne Tanke. Han opdagede nu, at den Leder, som forbinder de to modsatte Poler af den galvaniske Kjæde, og hvori al den Virkning, som ved noget Slags Electrometer kan opdages, er forsvunden, udøver en mægtig Sidevirkning, hvorved den sætter Magnetnaalen i Bevægelse.

Meddelelse om Electromagnetismens Opdagelse, Videnskabernes Selskabs Oversigter 1820–21

Dans la séance du 30 Octobre 1820, MM. Biot et Savart ont lu, à l'Académie des Sciences, un Mémoire dont l'objet est de déterminer, par des mesures précises, les lois physiques suivant lesquelles les fils de métal mis en communication avec les deux poles de l'appareil voltaïque agissent sur les corps aimantés.

A l'aide de ces procédés, MM. Biot et Savart ont été conduits au résultat suivant qui exprime rigoureusement l'action éprouvée par une molécule de magnétisme austral ou boréal placée à une distance quelconque d'un fil cylindrique très-fin et indéfini, rendu magnétique par le courant voltaïque. Par le point où réside cette molécule, menez une perpendiculaire à l'axe du fil: la force qui sollicite la molécule est perpendiculaire à cette ligne et à l'axe de fil. Son intensité est réciproque à la simple distance.

From the report published in *Annales de Chimie et de Physique XV* 1820

Pour avoir l'action d'une des parties du courant considérée isolément, il faut chercher suivant quelle loi les élémens du fil conducteur doivent agir pour que l'effet produit par leur réunion soit en raison inverse de la simple distance. M. de La Place a cherché cette loi par le calcul, et il a trouvé que si toutes les parties d'un fil rectiligne indéfini agissent en raison inverse du carré de leur distance à un point, leur action totale sera en raison inverse de la simple distance de ce point au fil conducteur. Cette belle expérience nous fournit donc la démonstration de l'hypothèse adoptée par M. Ampère dès le commencement de ses recherches, que chaque partie très-petite d'un courant agit sur une autre partie aussi très-petite en raison inverse du carré de la distance qui les sépare.

André Marie Ampère et Jacques Babinet *Exposé des nouvelles découvertes sur l'electricité et le magnetisme* Paris 1822

Lorsque deux conducteurs, ou plutót deux portions d'un méme conducteur voltaïque, l'une fixe et l'autre mobile, sont à une distance convenable dans des directions à peu pres parallèles, la portion mobile est attirée ou repoussée par la portion fixe, selon que la direction du courant électrique est la méme, ou en sens opposé dans ces deux portions.

André Marie Ampère *Annales de Chimie et de Physique* **18** 1821

Mon but, dans les cinq premiers paragraphes de ce Mémoire, les seuls dont je me propose de présenter ici une rapide analyse, est de déduire de la formule que j'ai donnée pour représenter l'action de deux portions infiniment petites de courans électriques, la valeur de l'action qui en résulte:

3°. Entre un élément et un systéme de courans circulaires d'un très-petit diamètre, dont les plans soient partout perpendiculaires à une ligne droite ou courbe passant par les centres des circonférences que les courans décrivent. C'est cette sorte de système, dont la forme est celle de la surface qu'on nomme ordinairement *surface canal*, que j'ai cru devoir désigner sous le nom de *solénoïde*, du mot grec σωληνοειδηζ, dérivé de σωληυ, canal, et qui signifie précisément qui a la forme d'un canal.

ANDRÉ MARIE AMPÈRE *Annales de Chimie et de Physique* **26** 1824

3.1. The basic equations

IN this chapter we shall be concerned with phenomena in which the main role is played by the current of charged particles. All our variables can be functions of space but are independent of time.

It is easy to present the relevant equations; we have all of eqns (1.1)–(1.7) but have to substitute $\partial/\partial t = 0$, yielding

$$\nabla \times \mathbf{H} = \mathbf{J}, \tag{3.1}$$

$$\nabla \times \mathbf{E} = 0, \tag{3.2}$$

$$\nabla \cdot \mathbf{D} = \rho, \tag{3.3}$$

$$\nabla \cdot \mathbf{B} = 0, \tag{3.4}$$

$$\mathbf{D} = \varepsilon \mathbf{E}, \tag{3.5}$$

$$\mathbf{B} = \mu \mathbf{H}, \tag{3.6}$$

$$\mathbf{F} = q(\mathbf{E} + \mathbf{v} \times \mathbf{B}). \tag{3.7}$$

There is a difficulty with eqn (3.6). It is correct for most of the chapter with μ taken as a constant but breaks down for ferromagnetic materials, which will be discussed in Section 3.11. All the other equations are all right but will not necessarily provide the simplest starting point for solving a given problem. We shall, therefore, introduce a number of alternative formulations.

First, we could search for some analogue of the potential function which proved so useful in electrostatics. We found there that by choosing a scalar function in the form $\mathbf{E} = -\nabla\phi$ we could automatically satisfy the other equation for the electric field strength, $\nabla \times \mathbf{E} = 0$. Can we do the same thing for the magnetic quantities? No, but we can do a similar thing. Since it is the equation $\nabla \cdot \mathbf{B} = 0$ that needs to be satisfied, we should choose the potential as a vector (called, not without logic, the *vector potential*), defined by the equation

$$\nabla \times \mathbf{A} = \mathbf{B}. \tag{3.8}$$

Substituting the above equation into eqn (3.1) we get

$$\nabla \times (\nabla \times \mathbf{A}) = \mu \mathbf{J}, \tag{3.9}$$

or using the vector relation (eqn (A.6)) we obtain the modified form

$$\nabla(\nabla \cdot \mathbf{A}) - \nabla^2 \mathbf{A} = \mu \mathbf{J}. \tag{3.10}$$

This equation can be further simplified to

$$\nabla^2 \mathbf{A} = -\mu \mathbf{J} \tag{3.11}$$

by choosing (in the physicist's jargon this is called choosing the gauge)

$$\nabla \cdot \mathbf{A} = 0. \tag{3.12}$$

Can we do that? One can't usually assign some arbitrary value to the divergence of a vector function. In the present case, however, we do have some freedom of choice. The definition of **A** by eqn (3.8) is not unique. If we add to **A** the gradient of a scalar function, the resulting vector $\mathbf{A}'$ still gives the same magnetic field because

$$\nabla \times \mathbf{A}' = \nabla \times (\mathbf{A} + \nabla \psi) = \nabla \times \mathbf{A} = \mathbf{B}. \tag{3.13}$$

Hence a suitable choice of ψ will ensure that

$$\nabla \cdot \mathbf{A}' = \nabla \cdot \mathbf{A} + \Delta \psi = 0. \tag{3.14}$$

So we are left with eqn (3.11). If the current density is specified, eqn (3.11) will provide the solution for **A** from which **B** can be determined. How can we find a solution for **A**? There is nothing easier. We only need to remember the differential equation for the scalar potential ϕ (eqn (2.12)) and its solution in the form of eqn (2.24). By analogy the general solution of eqn (3.11) is

$$\mathbf{A} = \frac{\mu}{4\pi} \int \frac{\mathbf{J}}{r} \, d\tau. \tag{3.15}$$

A simple integration will yield **A** if **J** is given. We shall go through a number of examples later. For the moment let us use the above expression for deriving Biot–Savart's law.

First we shall assume that the current density is confined to a thin wire in which case the integration variable may be changed to

$$d\tau = \mathbf{S} \cdot d\mathbf{l}, \tag{3.16}$$

where **S** is a vector normal to the cross-section, $|\mathbf{S}| = S$ is the area of the cross-section, and $d\mathbf{l}$ is an elementary vector along the tangent of the wire. Noting further that $\mathbf{J} \cdot \mathbf{S} = I$ and that I must be constant along the wire we may write eqn (3.15) in the modified form

$$\mathbf{A} = \frac{\mu I}{4\pi} \int \frac{d\mathbf{l}}{r}, \tag{3.17}$$

whence the magnetic field strength is

$$\mathbf{H} = \frac{\mathbf{B}}{\mu} = \frac{1}{\mu}\nabla\times\mathbf{A} = \nabla\times\frac{I}{4\pi}\int\frac{\mathbf{dl}}{r}. \tag{3.18}$$

We have to stop here for a moment to sort out the coordinates. As may be seen in Fig. 3.1 the coordinates of the wire element are x', y', z', whereas the coordinates of the point where we wish to determine the magnetic field are

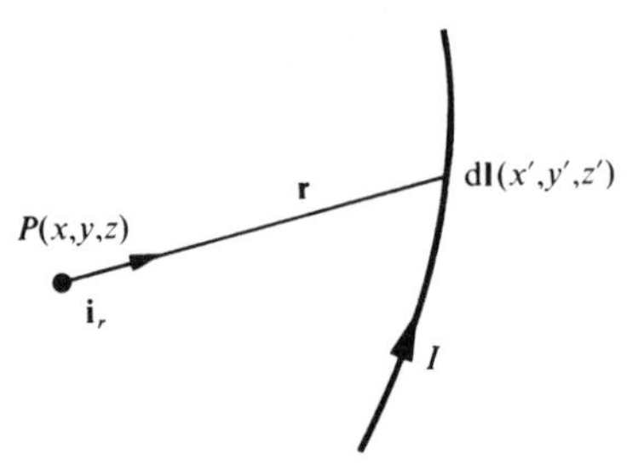

FIG. 3.1. A current element located at x', y', z' produces a magnetic field at x, y, z.

set of coordinates. Thus the curl operates on the coordinates of P but not on those of **dl** leading to

$$\mathbf{H} = \frac{I}{4\pi}\int\nabla\times\frac{\mathbf{dl}}{r} = \frac{I}{4\pi}\int\left(\nabla\frac{1}{r}\right)\times\mathbf{dl} = -\frac{I}{4\pi}\int\frac{(-\mathbf{i}_r)}{r^2}\times\mathbf{dl}$$

$$= -\frac{I}{4\pi}\int\frac{\mathbf{dl}\times\mathbf{i}_r}{r^2} \tag{3.19}$$

where $\mathbf{i}_r$ is the unit vector in the direction $\mathbf{r}$ and we have made use of the vector relation (A3) in the Appendix. This is Biot–Savart's law derived directly from Maxwell's equations.

Are there any other ways of determining the magnetic field? Well, one can use eqn (3.1) as it is, but very often one is better off by using its integral form that can be obtained by integrating both sides of eqn (3.1) over a surface

$$\int\nabla\times\mathbf{H}\,.\,\mathbf{dS} = \int\mathbf{J}\,.\,\mathbf{dS}. \tag{3.20}$$

By using Stokes' theorem for the left-hand-side and recognizing that the integral of the current density gives the current, the above equation reduces to

$$\oint_C\mathbf{H}\,.\,\mathbf{ds} = I, \tag{3.21}$$

where the line integration is along the curve C enclosing the surface. The positive sense of the integration path is defined relative to the positive current direction according to the usual right-hand convention. Equation (3.21) is known as Ampere's law.

We have now a number of equations describing the same thing. Which equation should we use in a practical case, the equation for the vector potential, Ampere's law, Biot–Savart's law, or attack directly Maxwell's equations? It's hard to tell. Experience with practical calculations helps, but even after years of work one usually remains this side of infallibility. I regret to say that there are no general guidelines available. If one manages to find the simplest method at the first attempt that can mostly be attributed to good luck. Can we claim then that the vector potential is as useful as its scalar counterpart in electrostatics? For computational purposes the answer is an unambiguous no. It is always laborious to find the components of a vector, so we are not much better off with **A** than with **H** or **B**. It turns out however that **A** is a more basic quantity of physics than **B**. Since **B** is given by the curl of **A** it is possible that **A** is finite while its curl is zero. Interestingly, under these conditions **A** has an effect on certain quantum-mechanical phenomena.† It would be unfair both to the vector potential and to quantum mechanics to say that none of those formulations have engineering applications (in fact the most sensitive magnetometer built to date is based on that kind of theory), but by and large engineers wouldn't lose much sleep if the use of **A** were banned with immediate effect. The vector potential is not a popular variable among engineers, maybe because it is not directly measurable. There are no instruments capable to measure the magnitude or direction of **A**.

In my opinion **A** is a useful thing if used with moderation. It helped us to derive Biot–Savart's law, it will come handy later in solving certain radiation problems, and it often leads to nice formulae, e.g. for the magnetic flux crossing a surface, defined as

$$\Phi = \int \mathbf{B} \cdot \mathrm{d}\mathbf{S} \tag{3.22}$$

that may be rewritten in terms of the vector potential as follows:

$$\Phi = \int \nabla \times \mathbf{A} \cdot \mathrm{d}\mathbf{S} = \oint_C \mathbf{A} \cdot \mathrm{d}\mathbf{s} \tag{3.23}$$

where C is the curve enclosing the surface.

We have now collected a good number of formulae which will serve us well in the following sections.

Before going on, just a few words about the classification of steady currents. It may be roughly divided into two parts: magnetostatics and the rest. We shall discuss the rest first (Sections 3.2–3.5), and that will give us some idea of the relative significance of electric and magnetic fields. Sections 3.6–3.16 are concerned with magnetostatics, where electric fields are assumed to be zero and the interrelationship of **J**, **H**, and **B** are studied.

Since we have so many variables, and since in each physical configuration only some of them appear, we will record (just for this chapter) the non-zero variables in each section heading.

† *Feynman lectures on physics* (1964). Vol. II, Section 15.5. Addison-Wesley, New York.

3.2. The defocusing of an electron beam (J, E, D, ρ, H, B)

We have so far talked about positive and negative charges, about point charges, and distributed charge. I shall now introduce the concept of an elementary charge, $1{\cdot}6\times10^{-19}$ C, carried by an elementary particle called the electron. This is not a necessity. It is possible to study electromagnetic theory without ever mentioning the word electron, but since it has become such a household word and is used so often we can just as well make use of it.

In the present section we shall consider a cylindrical electron beam of radius a in which the charge density is uniform ($\rho=\rho_0$) and all electrons travel with velocity $\mathbf{v}$. We are not going to enquire into the details how such beams can be produced (it belongs to the subject of physical electronics); we shall accept the fact that the beam exists and will try to work out the forces on the outermost electrons.

There is now cylindrical symmetry and a net charge per unit length

$$q=\pi a^2\rho_0, \tag{3.24}$$

yielding for the radial component of the electric field

$$E_R(a)=\frac{q}{2\pi\varepsilon_0 a}. \tag{3.25}$$

The magnetic field may be obtained from Ampere's law (eqn (3.21)) as follows

$$\int \mathbf{H}\,.\,\mathrm{d}\mathbf{s}=2\pi aH_\varphi=I, \tag{3.26}$$

where the line integral is taken over the circle of radius a (Fig. 3.2), I is the total current of the beam and H_φ is the azimuthal component of the magnetic field in the cylindrical coordinate system specified by R, φ, and z. From eqn (3.26) we get

$$H_\varphi=I/2\pi a. \tag{3.27}$$

The electric force on an electron of charge e travelling at the edge of the beam ($R=a$) is

$$\mathbf{F}=eE_R\mathbf{i}_R; \tag{3.28}$$

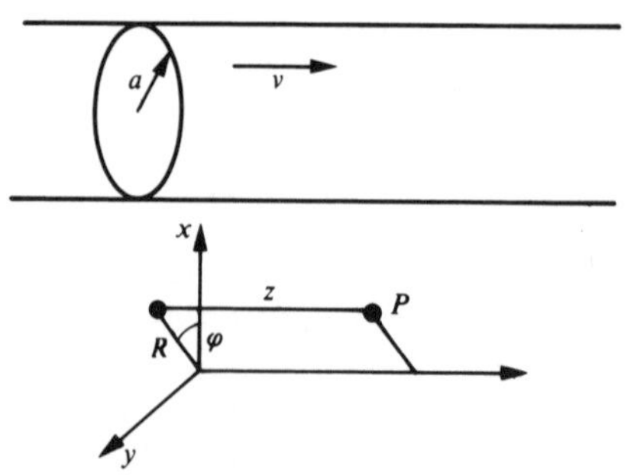

FIG. 3.2. An electron beam of radius a.

it is in the radial direction pointing outwards. The magnetic force is perpendicular both to the direction of motion ($+z$ axis) and the direction of magnetic field ($+\varphi$ direction). The vectorial product $\mathbf{v}\times\mathbf{B}$ gives an inward force in the radial direction. Hence the net force on the electron is

$$F = e(E_R - vB_\varphi) = \frac{e}{2\pi a}\left(\frac{q}{\varepsilon_0} - \mu_0 vI\right). \tag{3.29}$$

Noting that†

$$I = \rho_0 v a^2 \pi = qv \tag{3.30}$$

and anticipating the relationship

$$(\varepsilon_0\mu_0)^{-\frac{1}{2}} = c \tag{3.31}$$

to be derived in Section 5.7, we get the following formula for the force

$$F = \frac{eq}{2\pi a\varepsilon_0}\left(1 - \frac{v^2}{c^2}\right), \tag{3.32}$$

where c is the velocity of light. It may be seen from the above equation that the magnetic force is negligible in comparison with the electric force unless the velocity of the electron approaches the velocity of light. Thus the conclusion is that, owing to the repulsive forces between the electrons, a cylindrical electron beam is unstable. Using a technical term borrowed from electron optics we could say that the electron beam gets defocused.

The other conclusion (as far as it is permissible to generalize from a single example) is more important in principle. It is concerned with the relative magnitudes of electric and magnetic forces. Since charged particles rarely travel close to the velocity of light we may conclude that the magnetic forces are by orders of magnitude smaller than the electric forces. Why is it then that we have no difficulties in practice in observing magnetic fields? The reason is, of course, that there are two kinds of electric charges and their effect usually cancels, so the small magnetic field has a chance of getting observed.‡ And that brings us to the next example.

3.3. Pinch effect (J, H, B)

In contrast to our previous example we shall now investigate a cylindrical beam consisting of two kinds of charge carriers: negative electrons and some positive particles, which I do not wish to be more precise about at the moment. We shall assume that the two kinds of particles have equal densities and move in opposite directions. As a result there is no net space charge and

† Follows from the definition of current as the amount of charge crossing a given surface per unit time.

‡ We have not talked about relativity yet but you know that relativistic effects become significant when the particle velocity approaches the velocity of light. So we could, quite legitimately, regard magnetism as a relativistic effect.

hence no electric field. The currents, on the other hand, do not cancel because they represent opposite charges moving in opposite directions (remember, minus one times minus one is equal to plus one).

The magnetic field strength is given again by eqn (3.27); we only need to substitute for the current $I_n + I_p$ where the subscripts n and p refer to negative and positive particles respectively. Hence the magnetic force on the electron at the edge of the beam (at $R = a$) is

$$F_n = -ev_n B = -\frac{ev_n \mu_0}{2\pi a}(I_n + I_p), \tag{3.33}$$

and on the positive particle is

$$F_p = -ev_p B = -\frac{ev_p \mu_0}{2\pi a}(I_n + I_p). \tag{3.34}$$

Note that the force is inwards, and in the same direction for both particles.

Let us distinguish two cases.

Both particles are mobile†

In that case both of them will move inwards under the effect of magnetic force. But if the diameter of the beam is reduced, eqns (3.33) and (3.34) tell us that the forces are even larger. What happens then? If everything was uniform then the beam diameter would go on decreasing. In practice however the beam is not uniform and does not possess perfect cylindrical symmetry. Under these conditions the motion of the particles is fairly complicated, and of course our model is unable to predict the detailed behaviour of the particles. Nevertheless a few qualitative conclusions may be drawn without doing any further mathematics. If the cross-section happens to be smaller at a certain place, then the forces are larger there than at the neighbouring cross-sections, so the beam will be further constricted, etc., leading to the so-called sausage instability (Fig. 3.3(a)). If the magnetic field happens to be larger at one side than at the other side, then the beam will be deflected towards the weaker field which makes the field even weaker, etc., leading to the so-called kink instability (Fig. 3.3(b)).

The positive particle is immobile

In practice this means that the positive ions are part of the crystal lattice. There is then no magnetic force on the ions, only on the electrons. So the electrons want to move inwards but cannot because the ions hold them back by virtue of their electrostatic attraction. Naturally, if the ions attract the electrons the converse is true as well, i.e. owing to the magnetic force on the electrons, there is also an inward force upon the ions. So the whole crystal

† This can happen in a plasma or in a solid with equal numbers of holes and electrons.

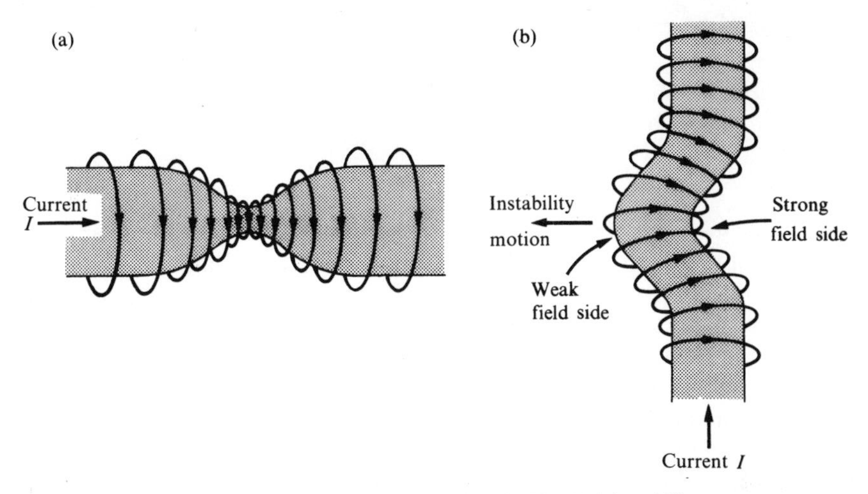

FIG. 3.3. (a) The sausage instability. (b) The kink instability.

structure tries to contract or in other words the material is under pressure (see Example 3.2). In any practical situation this pressure is small and can be neglected when calculating the forces on current-carrying conductors.

3.4. Flow patterns and Ohm's law (J, E, D)

Up to now we have not enquired into the question of how the current arose. We just assumed that certain charge carriers moved with certain velocities. Now we shall say that in a class of materials the current density at any point is proportional to the electric field; in mathematical form

$$\mathbf{J} = \sigma \mathbf{E}, \tag{3.35}$$

where σ is the conductivity of the material. The above equation in spite of its utmost simplicity represents an extremely good guess. It is valid for nearly all materials up to quite high electric fields.†

Will there be a magnetic field? Yes, of course, whenever there is a current there is a magnetic field as well. But the effect of the magnetic field upon the motion of the electrons is nearly always negligible (one exception causing the pinch effect is mentioned in the previous section). Thus we may safely ignore the magnetic field when determining the lines of current flow. We are then left with only one equation

$$\nabla \cdot \mathbf{E} = 0 \tag{3.36}$$

† The relationship breaks down when the material breaks down or when some esoteric phenomena become dominant, e.g. electron transfer between valleys of the conduction band—but we are trespassing again into the subject of the electric properties of materials.

or

$$\nabla . \mathbf{J} = 0 \tag{3.37}$$

or

$$\nabla^2 \phi = 0, \tag{3.38}$$

depending on our preference. Eqn (3.38) tells us that there is some analogy (see Example 3.3) with the electrostatic case treated in Chapter 2. We may, in fact, reinterpret any of the diagrams of Figs (2.25)–(2.31) by assuming that the whole space is filled with a material of conductivity σ and the field lines are now the lines of current flow as well.

Is the analogy perfect? Not really, because (1) as current flows there is some potential drop in the electrode itself, (2) zero conductivity for part of the space cannot be electrostatically modelled since there are no dielectrics with $\varepsilon_r = 0$, and (3) when current flows through two materials of different conductivity there is generally a surface charge at the boundary (see Example 3.4). It is not worth discussing any of these complications because one is rarely called upon to work out lines of current in a conductor. We shall rather return to a very simple geometrical configuration for deriving Ohm's law.

We shall take a piece of cylindrical material of length l and cross-section S and apply a voltage between the ends. The electric field is then

$$E = V/l, \tag{3.39}$$

the current density

$$J = \sigma V/l, \tag{3.40}$$

and the current

$$I = \sigma S/lV. \tag{3.41}$$

From circuit theory

$$V/I = R, \tag{3.42}$$

where R is the resistance of the material. Comparing eqns (3.41) and (3.42) we get

$$R = \frac{1}{\sigma} \frac{l}{S}. \tag{3.43}$$

Since $1/\sigma$ is by definition the resistivity we get the result you learned in school that the electrical resistance is proportional to the resistivity and the length of the sample and inversely proportional to its cross-section.

A more general definition of resistance valid for varying cross-sections may be easily arrived at, but it is hardly worth the trouble. The essential thing is that the relation $J=\sigma E$ is equivalent to Ohm's law. Risking the dismay of circuit engineers some theoreticians do, in fact, refer to eqn (3.35) as Ohm's law.

3.5. Electron flow between parallel plates (J, ρ, E, D)

We shall investigate here just one more physical configuration where in spite of having a finite current the magnetic field may be disregarded. The new feature will be a spatially varying space-charge density.

Let us take two parallel conducting plates in a vacuum, one of them endowed with the property that it is capable of emitting electrons. We shall further apply a voltage between the plates so that the electrons emitted by electrode 1 are attracted to electrode 2 (Fig. 3.4). The aim is to calculate the

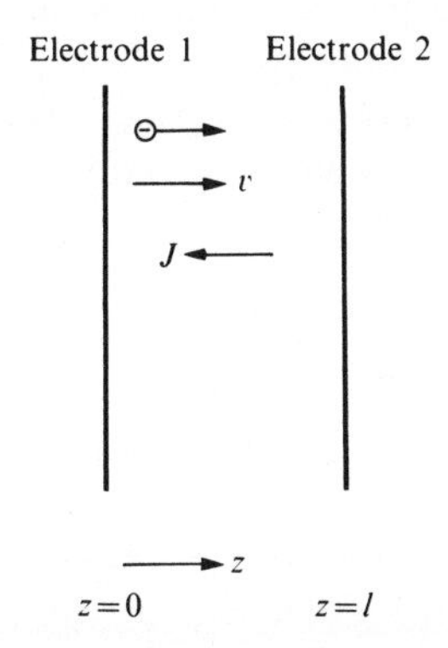

FIG. 3.4. Electron flow between two electrodes in a vacuum.

potential distribution between the plates, and the relationship between applied voltage and the magnitude of resulting current.

We can reduce the problem to a one-dimensional one by claiming that the plates are infinitely large. Alternatively, we may say that the plates are so close to each other (as it would be in a practical diode) that the electron beam hasn't got a chance to spread. Whichever way we look at it we can disregard with clear conscience both the radial electric field and the magnetic field. Hence the variables of interest are the current density, the electron velocity, the space-charge density, the longitudinal electric field strength, and the potential ϕ.

We can arbitrarily assign $\phi(0)=0$. Then $\phi(d)=V$, where V is the potential difference applied. The electron velocity at the point z may be worked out by the simple consideration that the electron gains kinetic energy at the expense of its potential energy (the same idea as in mechanics). Hence

$$\tfrac{1}{2}mv^2+e\phi=\tfrac{1}{2}mv^2(0)+e\phi(0). \tag{3.44}$$

Assuming now that the electrons are emitted with zero initial velocity, $\dot{v}(0)=0$, we get

$$v=\left(-\frac{2e}{m}\phi\right)^{\frac{1}{2}}. \tag{3.45}$$

Note that e is negative and that, for the above equation to apply, $\phi(z)>0$.†

The charge density may be obtained from the relationship

$$\rho=-\frac{J}{v}=-J\left(-\frac{2e}{m}\phi\right)^{-\frac{1}{2}}, \tag{3.46}$$

where J is a positive constant. ρ must of course be negative because it represents the space-charge density of electrons.

In this example the charges are in motion and the charge density varies from point to point but the density at a given point z is *not* dependent on time. Hence we are faced here with an electrostatic problem which may be solved with the aid of Poisson's equation (eqn (2.12)). Substituting into it the value of ρ from eqn (3.46) we get

$$\frac{\mathrm{d}^2\phi}{\mathrm{d}z^2}=\frac{J}{\varepsilon_0}\left(-\frac{2e}{m}\phi\right)^{-\frac{1}{2}}. \tag{3.47}$$

This is a reasonable-looking differential equation; I leave the solution to you (Example 3.5).

The main reason for showing this example is not its intrinsic value to applied scientists (the problem of space-charge-limited diodes is no longer in the forefront of interest) but to demonstrate the applicability of Poisson's equation under conditions of steady current flow.

3.6. The magnetic field due to line currents (J, H, B)

Assume that a current I flows along the z axis from minus infinity to plus infinity (Fig. 3.5(a)) and find the magnetic field at a distance R from the current. Owing to axial symmetry the magnetic field must be constant at a radius R, hence the application of Ampere's law yields

$$H2\pi R=I, \tag{3.48}$$

or

$$H=I/2\pi R. \tag{3.49}$$

In fact we have derived this relationship in Section 3.2. It makes no difference whether the current is distributed within a radius a or concentrated at the axis (only the enclosed current counts). Could we get the same result by using our formula for the vector potential (eqn (3.17))? We could,

† The basic assumption here is that an electron once started will necessarily reach the anode. The general case when electrons are permitted to turn back will not be treated here.

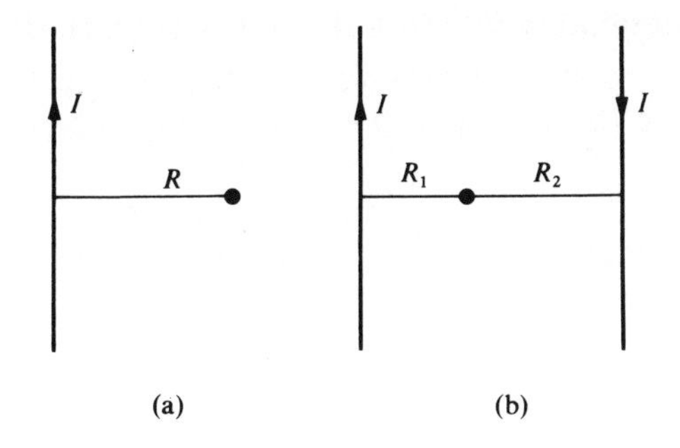

FIG. 3.5. Infinitely long line currents.

although we would run again into the problem of a diverging integral owing to the limits at infinity. We spent quite a long time sorting out this problem in the electrostatic case, and we need not repeat the argument here. Just by analogy with eqn (2.45) we obtain

$$A_z = \frac{\mu_0 I}{2\pi} \ln \frac{R_0}{R}. \tag{3.50}$$

For obtaining **B** we take the curl of the vector potential in a cylindrical coordinate system R, φ, z as follows

$$\nabla \times \mathbf{A} = \frac{1}{R} \begin{vmatrix} \mathbf{i}_R & R\mathbf{i}_\varphi & \mathbf{i}_z \\ \partial/\partial R & \partial/\partial \varphi & \partial/\partial z \\ 0 & 0 & A_z \end{vmatrix}. \tag{3.51}$$

Recognizing that nothing can change in the φ and z directions ($\partial/\partial\varphi = 0$, $\partial/\partial z = 0$), the only non-zero component is provided by

$$B_\varphi = -\frac{\partial A_z}{\partial R} = \frac{\mu_0 I}{2\pi R}, \tag{3.52}$$

in agreement with eqn (3.49).

For two line currents flowing in opposite directions (Fig. 3.5(b)), we may write Ampere's law twice and add the magnetic fields or add the vector potentials. In either case we obtain

$$\mathbf{H} = \frac{I}{2\pi}\left(\frac{\mathbf{i}_{\varphi 1}}{R_1} - \frac{\mathbf{i}_{\varphi 2}}{R_2}\right) \tag{3.53}$$

where $\mathbf{i}_{\varphi_1}$ and $\mathbf{i}_{\varphi_2}$ are unit vectors in the azimuthal directions from the two line currents respectively.

3.7. The magnetic field due to a ring current (J, H, B)

The solution for line currents was simple enough. Unfortunately the determination of the magnetic field for a ring current needs a lot of calculation.

We shall assume that a ring of radius a situated in the $z=0$ plane carries a current I (Fig. 3.6) and we wish to determine the magnetic field at the point

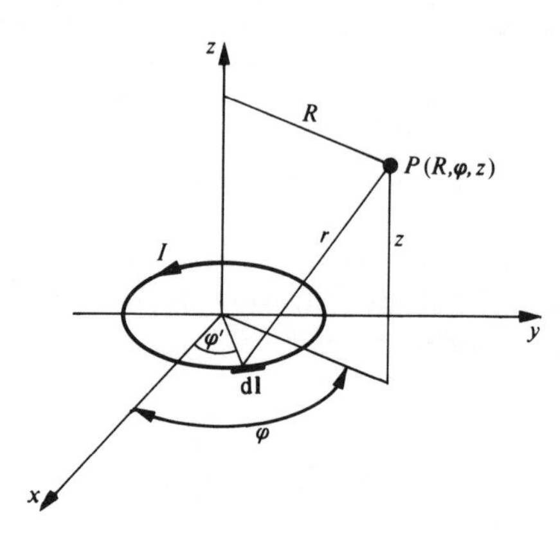

FIG. 3.6. Geometry for determining the vector potential due to a ring current.

$P(R, \varphi, z)$, i.e. at an arbitrary point in space. We shall rely on the vector-potential formulation because that appears to be best suited for utilizing the axial symmetry of the chosen geometry. The vector potential due to any current element is always in the direction of the current element. Hence if the current is everywhere in the azimuthal direction the vector potential must be in the same direction too. Thus even before starting any calculations we may immediately say that $\mathbf{A}$ will only have an A_φ component and that will be independent of φ.

Taking the coordinates of the current element $\mathbf{dl}$ as $a, \varphi', 0$ (Fig. 3.6) and noting that

$$\mathbf{dl} = a\,\mathrm{d}\varphi'(-\sin\varphi'\mathbf{i}_x + \cos\varphi'\mathbf{i}_y) \tag{3.54}$$

we obtain for the desired component of the vector potential

$$A_\varphi = \mathbf{A}\,.\,\mathbf{i}_\varphi = \frac{\mu_0 I}{4\pi}\int\frac{\mathbf{dl}\,.\,\mathbf{i}_\varphi}{r} = \frac{\mu_0 Ia}{4\pi}\int_{2\pi}\frac{\cos(\varphi'-\varphi)\,\mathrm{d}\varphi'}{r} \tag{3.55}$$

or

$$A_\varphi = \frac{\mu_0 Ia}{4\pi}\int_{2\pi}\frac{\cos(\varphi'-\varphi)\,\mathrm{d}\varphi'}{\{(R\cos\varphi - a\cos\varphi')^2 + (R\sin\varphi - a\sin\varphi')^2 + z^2\}^{\frac{1}{2}}}, \tag{3.56}$$

where the expression for r has been substituted.

Since A_φ must be independent of φ we can simplify eqn (3.56) by taking $\varphi = 0$. The remaining integration is nonetheless difficult, and not expressible in terms of simple functions. Mathematicians, wisely foreseeing such difficulties, worked out the theory of a number of special functions and tabulated them as well. In the age of the computer their work is no longer indispensable but is still useful as a sort of short-hand notation. If you look up the relevant books you will find that eqn (3.56) may be expressed in terms of *elliptic integrals.* Having obtained the vector potential the magnetic field may be obtained by the usual differentiation which, incidentally, also leads to elliptic integrals.

There is nothing fearful in elliptic integrals. Their definitions are fairly simple, they are nicely tabulated and they have a few interesting interrelations. All that is easily digestible. My main reason for not introducing elliptic integrals here (you can though attempt Examples 3.9–3.11 if you wish to have some experience in handling them) is that I do not want to burden your memory with new formulae, and besides, for cases of most practical interest (the field far away from the ring and in the vicinity of the axis) eqn (3.56) may be integrated out, as will be presently seen.

For both cases mentioned above ($R^2 + z^2 \gg a^2$ and $a^2 + z^2 \gg R^2$), the following inequality is valid

$$R^2 + z^2 + a^2 \gg 2aR \cos \varphi'. \tag{3.57}$$

Hence we can expand the denominator of the integrand as follows:

$$\begin{aligned}(R^2 + a^2 + z^2 - 2aR \cos \varphi')^{-\frac{1}{2}} &= (R^2 + a^2 + z^2)^{-\frac{1}{2}}\left(1 - \frac{2aR \cos \varphi'}{R^2 + a^2 + z^2}\right)^{-\frac{1}{2}} \\ &\cong (R^2 + a^2 + z^2)^{-\frac{1}{2}}\left(1 + \frac{aR \cos \varphi'}{R^2 + a^2 + z^2}\right).\end{aligned} \tag{3.58}$$

Substituting the above approximate relation into eqn (3.56) we can perform the integration, yielding

$$A_\varphi = \frac{\mu_0 I}{4} \frac{a^2 R}{(R^2 + a^2 + z^2)^{\frac{3}{2}}}. \tag{3.59}$$

Let us first consider the case when the point of observation is far away from the ring ($R^2 + z^2 \gg a^2$); then the vector potential takes the form

$$A_\varphi = \frac{\mu_0 I}{4} \frac{a^2 R}{(R^2 + z^2)^{\frac{3}{2}}}, \tag{3.60}$$

which may be rewritten in spherical coordinates as

$$A_\varphi = \frac{\mu_0 I a^2}{4} \frac{\sin \theta}{r^2}, \tag{3.61}$$

whence the magnetic field is

$$\mathbf{H}=\frac{1}{\mu_0}\nabla\times\mathbf{A}=\frac{1}{r^2\sin\theta}\begin{vmatrix}\mathbf{i}_r & r\mathbf{i}_\theta & r\sin\theta\mathbf{i}_\varphi\\ \partial/\partial r & \partial/\partial\theta & \partial/\partial\varphi\\ 0 & 0 & r\sin\theta A_\varphi\end{vmatrix}. \tag{3.62}$$

Noting that $\partial/\partial\varphi=0$ we get for the components of the magnetic field

$$\left.\begin{aligned}H_r&=\frac{Ia^2\pi}{4\pi}\frac{2\cos\theta}{r^3},\\ H_\theta&=\frac{Ia^2\pi}{4\pi}\frac{\sin\theta}{r^3},\\ H_\varphi&=0.\end{aligned}\right\} \tag{3.63}$$

Can you remember seeing these self-same components somewhere before? Well, the constants are different but apart from that eqn (2.33), the formula for the electric field of an electric dipole, looks the same. On the basis of this analogy we may call a ring current a magnetic dipole, or more precisely we should say that sufficiently far away from a ring current the magnetic field appears as if it was created by two closely spaced magnetic charges (which of course do not exist).

There is another less obvious conclusion at which one might arrive by inspecting eqn (3.63). Notice that the area of the ring πa^2 appears as one of the factors in the constant. It turns out (though we are not going to prove it here) that the exact shape of the current loop is immaterial (not unreasonable if the loop is far away) and in the general case we only need to replace πa^2 by S, the area of the loop.

Turning now to the case when the point of observation is in the vicinity of the axis, we get for the vector potential

$$A_\varphi=\frac{\mu_0 I}{4}\frac{a^2R}{(a^2+z^2)^{\frac{3}{2}}}, \tag{3.64}$$

and for the magnetic field (in cylindrical coordinates this time)

$$\left.\begin{aligned}H_R&=\frac{3I}{4}\frac{a^2Rz}{(a^2+z^2)^{\frac{5}{2}}},\\ H_\varphi&=0,\\ H_z&=\frac{I}{2}\frac{a^2}{(a^2+z^2)^{\frac{3}{2}}}.\end{aligned}\right\} \tag{3.65}$$

3.8. The magnetic field inside a solenoid (J, H, B)

A solenoid is a tightly wound coil. Each turn may be regarded equivalent to a ring in which a current I flows. For calculating the magnetic field we shall take the coordinate system shown in Fig. 3.7, where the z axis coincides with the axis of the solenoid. At an arbitrary point on the axis ($z=\zeta$, $R=0$) the

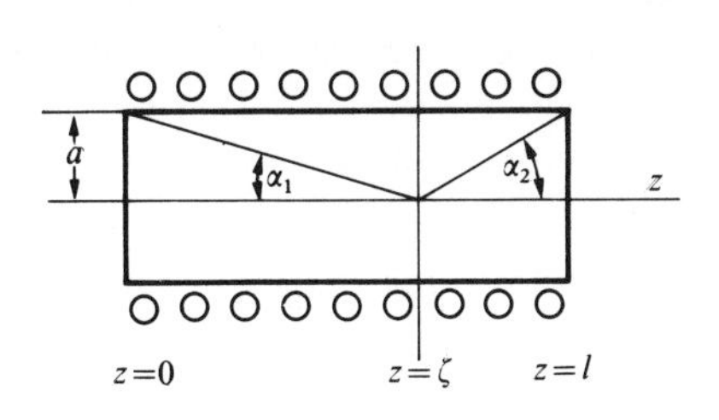

FIG. 3.7. A solenoid.

magnetic field may be obtained by summing the contribution of each turn. It is actually easier to do the calculation if instead of individual turns we assume that the current is continuously distributed on the surface of the cylinder, and work in terms of I_l, the current per unit length, obtained by dividing the total current (NI) by the length l of the solenoid, i.e.

$$I_l = NI/l. \tag{3.66}$$

The current flowing at z in the interval $\mathrm{d}z$ is then $I_l\,\mathrm{d}z$, at a distance $\zeta - z$ from the point where we wish to determine the magnetic field. According to eqn (3.65) such a current will create a magnetic field

$$\mathrm{d}H_z = \frac{I_l\,\mathrm{d}z}{2}\frac{a^2}{\{a^2+(\zeta-z)^2\}^{\frac{3}{2}}} \tag{3.67}$$

and the total magnetic field may be obtained by integrating over the length of the solenoid as follows:

$$H_z = \frac{I_l a^2}{2}\int_0^l \frac{\mathrm{d}z}{\{a^2+(\zeta-z)^2\}^{\frac{3}{2}}}. \tag{3.68}$$

Changing the integration variable to the angle α by the relationship

$$\cot\alpha = \frac{\zeta-z}{a}, \tag{3.69}$$

we may perform the integration and get at the end

$$H_z = \frac{NI}{2l}(\cos\alpha_1 + \cos\alpha_2). \tag{3.70}$$

If the solenoid is very long ($l \gg a$), then $\cos\alpha_1 \approx \cos\alpha_2 = 1$, and

$$H_z = NI/l. \tag{3.71}$$

This is of course not valid near the ends but if the solenoid is long enough, eqn (3.71) may be regarded valid for most of its length.

Incidentally, eqn (3.71) may be derived much more simply if we make the *a priori* assumption that the magnetic field is constant inside the solenoid and zero outside. We may then apply Ampere's law to the path shown in Fig. 3.8 yielding

$$Hl = NI. \tag{3.72}$$

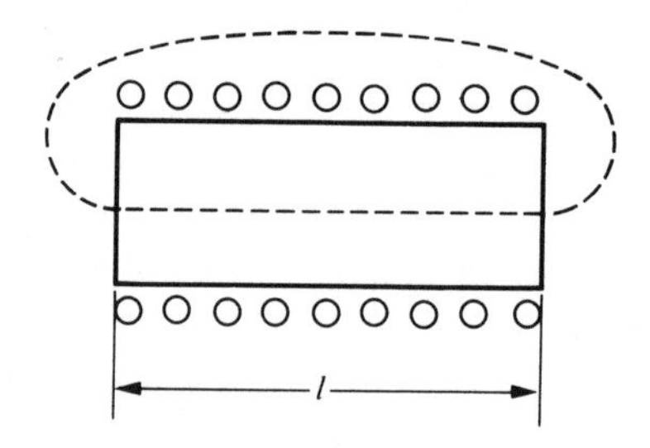

FIG. 3.8. A closed path for applying Ampere's law.

It is a bit of a coincidence that eqns (3.71) and (3.72) agree. The reason is that in the latter approach the magnetic field is overestimated near the ends inside the solenoid, and underestimated outside the solenoid. The two inaccuracies just about balance each other.

3.9. Further analogies with electrostatics

What happens in the electrostatic case if we fill the whole space by a dielectric? The electric field remains unchanged and the electric flux density increases by a factor ε_r. What happens in the 'steady current' case if we fill the whole space by a magnetic material? The magnetic field remains unchanged and the magnetic flux density increases by a factor μ_r.

Does the analogy still hold when the magnetic material fills only part of the space? The answer is not quite, because the boundary conditions are not quite the same. Using the technique (surface and line integrals shrinking to zero) as in Section 2.10 we get the boundary conditions

$$B_{n1} - B_{n2} = 0 \quad \text{and} \quad H_{t1} - H_{t2} = K, \tag{3.73}$$

where K is a surface current. Hence in the absence of surface charges and surface currents the analogy still holds.

If we look at the magnetic field in a region in which no currents flow, the analogy is even closer because $\nabla\times\mathbf{H}=0$ and we can introduce a magnetic scalar potential with the relation

$$\mathbf{H}=-\nabla\phi_{\mathrm{m}}. \tag{3.74}$$

A good example is the calculation of the effect of a spherical piece of magnetic material inserted into a homogeneous magnetic field. The mathematics is exactly the same as in the electrostatic case; we need only to replace E_0 by H_0 and ε_{r} by μ_{r} in eqn (2.110) in order to get ϕ_{m}. Hence Fig. 2.41 (p. 47) is a valid representation for magnetic materials as well. The difference between the two cases is merely quantitative. The magnetic materials used in practice have a μ_{r} several orders of magnitude higher than the ε_{r} of practical dielectric materials. Hence the 'attraction' of the field lines is much more pronounced in the magnetic case. To give a practical example let us put in the middle of our solenoid a magnetic material of high μ_{r} as shown in Fig. 3.9. Now practically all the flux lines are funnelled into the magnetic material so much so that we are entitled to regard both **H** and **B** as being zero outside the magnetic material.

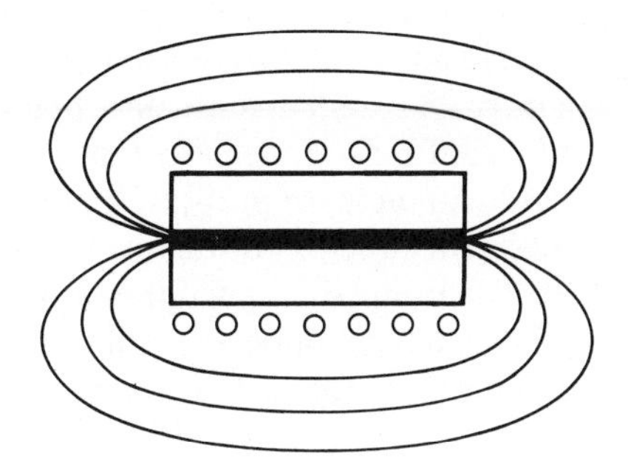

FIG. 3.9. Magnetic flux lines funnelled into the high-permeability core.

3.10. Magnetic materials

When studying the electromagnetic properties of dielectrics I was very reluctant to get involved with the physics of dielectrics. Now I am even more reluctant to talk about the physics of magnetic materials. Not so much because I don't understand the subject (that is no real obstacle to a lecturer), but more because of the time we are likely to consume, even if we keep a respectable distance from quantum mechanics and concentrate solely on phenomenological theories.

It must be admitted that magnetic materials behave most unreasonably. The relative permeability may turn out to be a tensor; it may take on very high values, say 10^6 or more, or it may have a value very close to unity, say 1·00002 for a wide range of temperatures, and then on cooling the material another few millidegrees μ_r may drop to zero.

Why to bother so much about the details, you may ask; couldn't we just say that $\mathbf{B} = \mu\mathbf{H}$ and go on considering further examples? The trouble is that for the most widely used magnetic material—iron—μ_r is not a constant. Not only that its value depends on **H**, but, even worse, it depends on the previous history of the sample. So we have to discuss the B–H curves, and I will include a few more things, but you must realize that there is a lot more to it. By the time we finish with it we shall have just scratched the surface of the subject.

3.11. The *B–H* curve of ferromagnetic materials

It is customary to divide all magnetic materials into diamagnetic ($\mu_r < 1$), paramagnetic ($\mu_r \gtrsim 1$), and ferromagnetic ($\mu_r \gg 1$) groups. Paramagnetic materials are unimportant† from an engineering point of view, and diamagnetic materials may have a future; the present, however, belongs to the ferromagnetic group, and above all to the most important representative of the group—the various alloys of iron.

In order to obtain the B–H curve let us make the following experiment. Place a cylindrical iron rod inside a solenoid (as in Fig. 3.9), vary the current and measure the flux density. We shall assume that, before we switch on the current, $H = 0$ and $B = 0$, a natural-enough assumption. As we increase the current, the magnetic field will increase proportionally (eqn (3.71)) but the flux density will be a nonlinear function of the magnetic field, as shown by the line OP in Fig. 3.10. The point P is called the saturation point, beyond which $\partial B/\partial H = \mu_0$, i.e. the iron has stopped contributing to the flux density; any further increase of H will result in that much increase of B as in a vacuum. Reducing now the current (and H with it) we shall not retrace the same curve; B will decrease much more slowly and will have a finite value at zero current. If we now increase the current in the opposite direction B will decrease further reaching zero at R, and negative saturation at S. The other half of the curve $STUP$ displays the same behaviour. This is no doubt a remarkable curve. B being so sluggish, it is usually referred to as the hysteresis curve.‡ Note that the B–H relationship is irreversible everywhere inside the hysteresis curve. If at an arbitrary point (say V) we decided to decrease the current, B would decrease in a different manner, as shown in Fig. 3.10.

† The paramagnetic maser is a rather special exception.

‡ Derived from the Greek word *hystereein* = to be behind.

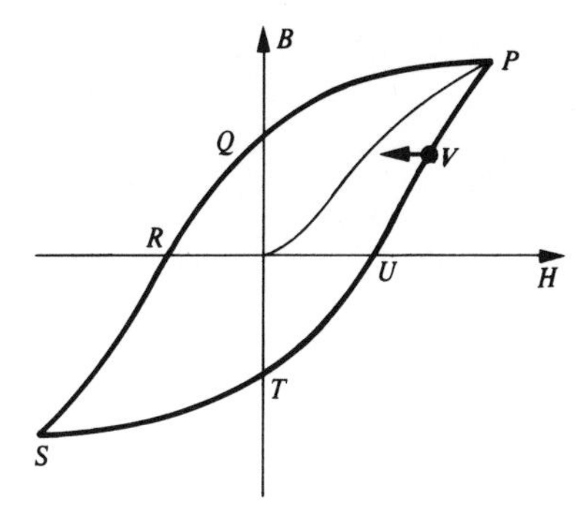

FIG. 3.10. The hysteresis curve.

As far as engineering applications are concerned the most remarkable feature of the curve is that we can get a flux density even in the absence of all external agents. This is of course the permanent magnet you have all come across.

How is this possible? Up to now a current appeared to be absolutely essential for creating a magnetic flux. The obvious way out of this dilemma is to say that the currents are there—inside the magnetic material. The detailed mechanism of these currents has kept lots of physicists busy ever since Ampère made the hypothesis, but apparently some more time is needed to find a proper solution.

For permanent magnets a wide hysteresis curve is needed so that demagnetization should not easily occur. For transformers and rotating machinery a narrow hysteresis curve is preferable because irreversibility leads to losses.

I am afraid this is about as much as I am going to say about magnetic materials apart from a very brief discussion of some pseudomagnetic materials (superconductors) in Sections 3.15 and 3.16. We shall look at a few examples now in which simple geometries will be considered.

3.12. The magnetic flux density inside a permanent magnet of toroidal shape (B)

In a permanent magnet shaped as a torus we have $H=0$. The only equation to satisfy is

$$\nabla \cdot \mathbf{B}=0, \tag{3.75}$$

and that means that all the flux lines must be closed. Choosing a toroidal shape there are still lots of possible ways for the lines to arrange themselves, but in a good permanent magnet the flux lines will be circles with centres at

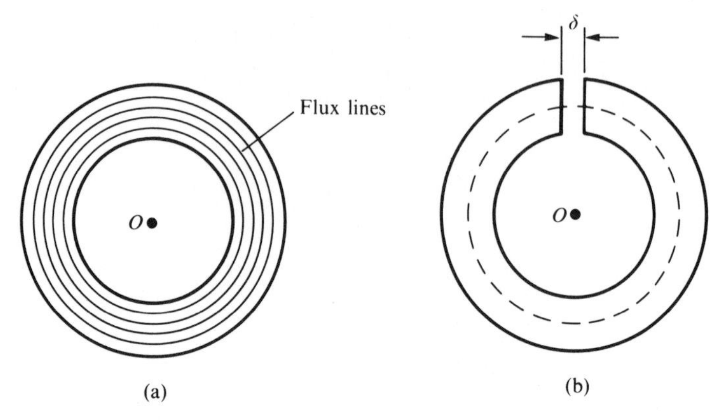

FIG. 3.11. (a) Magnetic field lines inside a permanent magnet. (b) The same magnet with a narrow gap.

O, the centre of the toroid (Fig. 3.11(a)). We may then claim that $B =$ constant everywhere inside the magnet and zero outside.

3.13. The magnetic field inside a permanent magnet of toroidal shape containing a gap (H, B)

In the last section we have come to the interesting conclusion that **B** may alone exist of all our variables but we reached that conclusion on a magnet shape not much used in practice. If we take the trouble to make a permanent magnet we would like to have access to the magnetic flux so let us look at the more practical case (Fig. 3.11(b)) when a narrow gap is cut into the magnet. How will the flux density vary in the gap? It will be hardly different from B_0, the value in the material; in the short space available the flux lines have not got a chance to spread. We shall therefore take the flux density in the gap to be equal to B_0. The corresponding magnetic field will be B_0/μ_0. What about the value of the magnetic field in the magnet? It may be obtained from the consideration that in the absence of an external current the line integral of H (taken over the dotted lines in Fig. 3.11(b)) must vanish, leading to

$$H_0\delta + H_i l = 0, \tag{3.76}$$

where δ is the width of the gap and l is the length of the path in the magnet. From eqn (3.76) we get the value of the magnetic field inside the magnet as

$$H_i = -\frac{B_0\delta}{\mu_0 l}. \tag{3.77}$$

There is still one question we have to ask, Will the magnetic flux density inside the material B_0 be the same as B_r the value before the gap was cut?

No, not quite. There are now two relations to satisfy. B_0 is obtained where the straight line

$$B = -\frac{\mu_0 l}{\delta} H \tag{3.78}$$

intersects the B–H hysteresis curve as shown in Fig. 3.12.

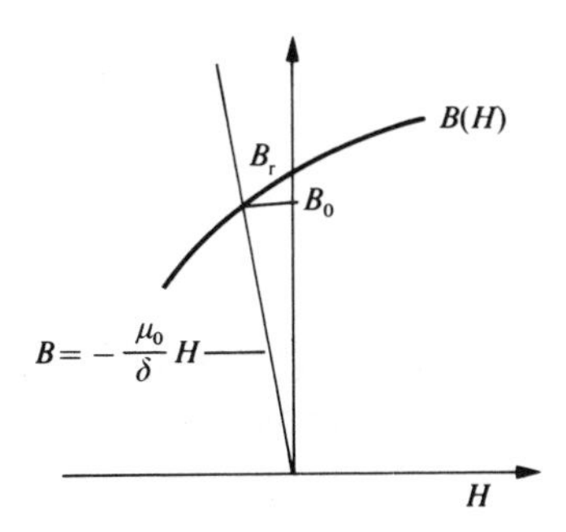

FIG. 3.12. Construction for finding B_0.

3.14. The magnetic field in a ferromagnetic material of toroidal shape excited by a steady current (J, H, B)

We shall now take a ferromagnetic material that has a very narrow hysteresis loop so that we can assume with good approximation a unique (though of course nonlinear) relationship between B and H. The material is again assumed to be of a toroidal shape but it is now excited by a current I flowing in a coil of N turns (Fig. 3.13(a)). We may then apply Ampère's law to the path shown by dotted lines to obtain

$$H_1 l = NI, \tag{3.79}$$

whence

$$H_1 = NI/l, \tag{3.80}$$

—a formula we have already met (eqn (3.71)) when discussing an approximate solution for a long solenoid. The corresponding value of the magnetic flux density is $B_1 = B(H_1)$ that can be obtained from the B–H curve, as shown in Fig. 3.14.

Next, we shall find the solution when there is a gap of width δ in the material (Fig. 3.13(b)). For the same current the magnetic flux density will decrease to B_2, its value being the same both in air and in the material. Denoting the magnetic field in the material by H_{2i} and in air by H_{20}, Ampère's law yields

$$H_{20}\delta + H_{2i} l = NI. \tag{3.81}$$

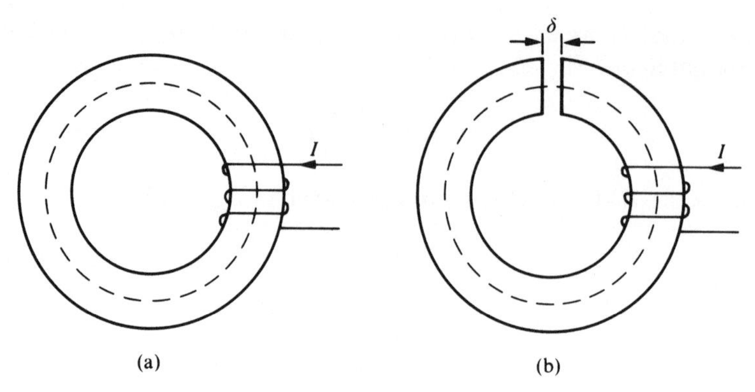

FIG. 3.13. Soft iron core magnetized by a current I with and without gap.

Noting further that $H_{20} = B_2/\mu_0$, the above equation gives a linear relationship between H_{2i} and B_2. Hence the solution is obtained by the intersection of the straight line

$$B = \frac{\mu_0 NI}{\delta} - \mu_0 H \frac{l}{\delta} \tag{3.82}$$

and the B–H curve. The graphical construction is shown in Fig. 3.14. If we wish to produce the same flux density as in the absence of the gap, we need to increase the current. This is the subject of Example 3.13.

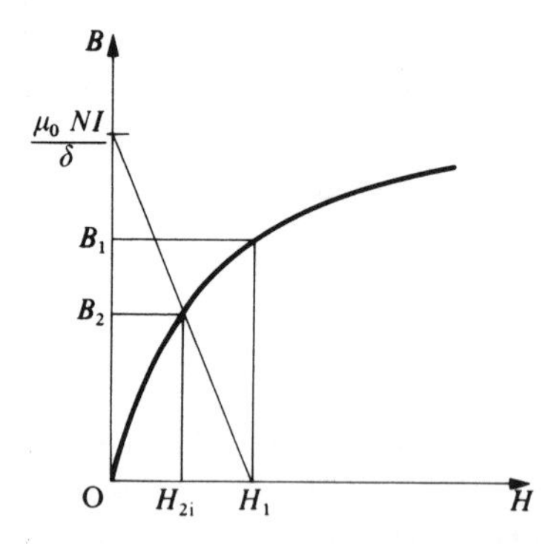

FIG. 3.14. Construction for finding B_2.

3.15. The perfect diamagnet (H, B, K)

There is a class of materials called type I superconductors† which below a certain critical temperature (around the normal boiling-point of liquid

† As you may have guessed there exist also type II superconductors with intriguing magnetic properties, but I can't talk about them now; that is really beyond the scope of the present lectures.

helium ≈ 4 K) become both perfect conductors and perfect diamagnets. This means that **B** must be zero inside the material. Thus if we place a piece of such material (say a sphere) into an otherwise constant magnetic flux, the material will expel the flux lines as shown in Fig. 3.15. How can this happen? What mechanism is responsible for expelling the magnetic flux? The appearance of surface currents (which we have previously denoted by **K**). As the sphere becomes superconducting, surface currents are set up producing a magnetic flux opposite to that already existing inside the material.

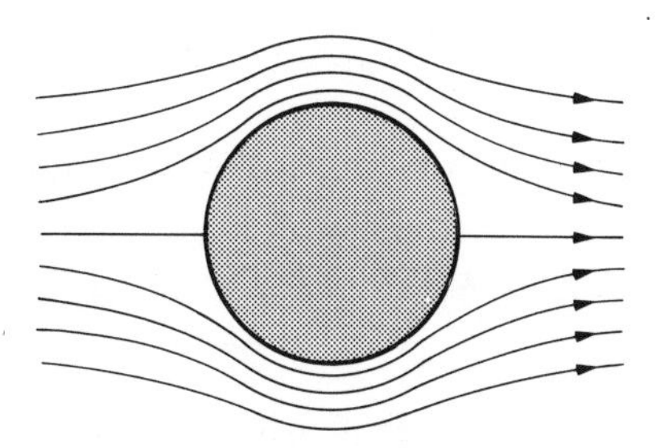

FIG. 3.15. A perfectly diamagnetic sphere expels the magnetic field lines.

3.16. The penetration of the magnetic flux into a type I superconductor (J, H, B)

In real life nothing is ever perfect.† A type I superconductor, I confess, is not a perfect diamagnet; it will let in the magnetic flux just a little bit. How far will the magnetic flux penetrate? Is there a simple way of describing the decay of the magnetic flux mathematically? There is. We can use the following one-dimensional model.

Half of the space ($z < 0$) is a vacuum in which a constant flux density $\mathbf{B} = B_0\mathbf{i}_y$ is assumed. The corresponding vector potential (assuming no variation in the transverse direction $\partial/\partial x = \partial/\partial y = 0$) is

$$\mathbf{A} = (A_0 + A_1 z)\mathbf{i}_x, \tag{3.83}$$

where A_0 and A_1 are constants. The other half of the space ($z > 0$) is filled with a type I superconductor.

The approach I am going to adopt now applies to the present section only and acknowledges the fact that a superconductor is not an 'ordinary' magnetic material; it cannot be described by a 'magnetic' constant, by assigning to it a certain value for μ_r. Instead, a new macroscopic constant needs to be introduced.

† The 'perfect conductor' aspect of a superconductor might be an exception; the resistance is so small that no one has so far been able to measure it.

We shall proceed similarly as in the case of conductors. We assumed then that the electric field was proportional to the current density. Now we shall assume that there is a linear relationship between the vector potential and the current density

$$\mathbf{J} = -\gamma \mathbf{A}, \tag{3.84}$$

where γ is the new macroscopic constant. Substituting eqn (3.84) into (3.11) and retaining the one-dimensional character of the problem, we get the differential equation

$$\frac{\partial^2 A_x}{\partial z^2} - \mu_0 \gamma A_x = 0, \tag{3.85}$$

the relevant solution of which (disregarding the exponentially increasing term) is as follows:

$$A_x = A_2 \, \mathrm{e}^{-\kappa z}, \qquad \kappa = (\mu_0 \gamma)^{\frac{1}{2}}. \tag{3.86}$$

where A_2 is another constant.
Hence we get for the magnetic flux density

$$B_y = \frac{\partial A_x}{\partial z} = -\kappa A_2 \, \mathrm{e}^{-\kappa z}. \tag{3.87}$$

The flux density reduces to $1/\mathrm{e}$ of its value at a distance $\lambda = 1/\kappa$, which is called the penetration depth. A typical value is $\lambda = 60$ nm (nm = nanometre), so the material is not very far from being a perfect diamagnet.

3.17. Forces

Let us first work out the force upon a current element of length $\mathrm{d}s$ in the presence of a flux density $\mathbf{B}$. According to eqn (3.7) the force on a point charge q is

$$\mathbf{F} = q(\mathbf{v} \times \mathbf{B}). \tag{3.88}$$

Assuming now that the magnetic field is constant over S_0, the cross-section of the wire, the force upon all charges within the volume element $S_0 \, \mathrm{d}s$ must be the same. Hence the total force on the current element is

$$\mathbf{F} = \rho S_0 v \, \mathrm{d}\mathbf{s} \times \mathbf{B} = I \, \mathrm{d}\mathbf{s} \times \mathbf{B}, \tag{3.89}$$

and for a whole current loop we get

$$\mathbf{F} = I \oint (\mathrm{d}\mathbf{s} \times \mathbf{B}). \tag{3.90}$$

If the magnetic flux density is constant over the whole loop then $\mathbf{B}$ may be taken out of the integral sign, and the remaining integration yields zero.

Consequently, there is no net force upon a current loop in a homogeneous magnetic flux. There is, however, a torque which may be easily calculated for a rectangular loop. In the specific example of Fig. 3.16 (your guess is correct if you think that the arrangement has something to do with electrical machines) there is a current-carrying loop capable to rotate around the horizontal axis in the magnetic flux (assumed constant) of the permanent magnet. As may be seen from Fig. 3.16(b) the forces on sides 1 and 3 balance

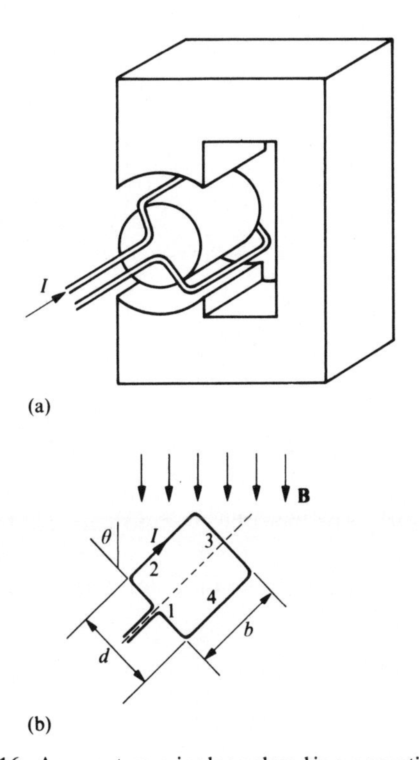

FIG. 3.16. A current-carrying loop placed in a magnetic field.

each other, whereas those acting upon sides 2 and 4 produce a torque. The force on side 2 is

$$\mathbf{F} = I\left|\int_0^b (\mathrm{d}\mathbf{s} \times \mathbf{B})\right| = IbB \tag{3.91}$$

whence the torque comes to

$$T = IbdB \cos\theta \tag{3.92}$$

where θ is the angle between the plane of the loop and the vertical direction.

Next we shall work out the force upon a small current loop due to another current loop in the geometry of Fig. 3.17. Owing to symmetry the magnetic field of loop 1 has only H_z and H_R components. In the vicinity of the z axis they are given by eqn (3.65). We shall consider the forces due to each component separately.† The force due to H_z upon the current element in loop 2 is in the radial direction. Integrating over all current elements the net

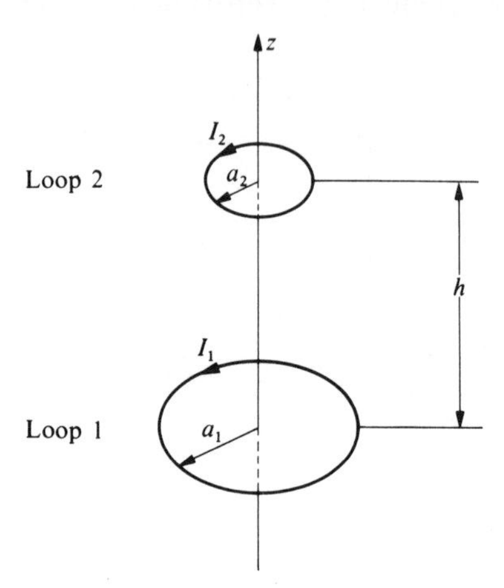

FIG. 3.17. Two current-carrying loops.

force is obviously zero. For the other component, H_R the force $\mathbf{i}_\varphi \times \mathbf{i}_R$ is in the negative z direction. (Remember from school? 'The force between two currents flowing in the same direction is attractive.') Integration over the circumference yields

$$\mathbf{F} = -\mathbf{i}_z \mu_0 I_2 H_R a_2 \int \mathrm{d}\varphi$$

$$= -\frac{3\pi}{2} \frac{\mu_0 I_1 I_2 a_1^2 a_2^2 h}{(a_1 + h^2)^{\frac{5}{2}}} \mathbf{i}_z. \qquad (3.93)$$

Examples 3

3.1. A current with constant current density J_0 flows in the z-direction between the cylinders $x^2 + y^2 = a^2$ and $x^2 + y^2 = b^2$.

Determine the magnetic field in the regions

(i) $r < a$, (ii) $a < r < b$, (iii) $r > b$.

3.2. A cylindrical column of mercury 10 mm diameter carries a current of 100 A uniformly distributed over the cross-section. Calculate the pressure due to the pinch effect, (i) at a radius of 2·5 mm and (ii) at the axis of the conductor.

† This is, of course, not necessary. We could just substitute eqn (3.65) into (3.90) and perform the integration.

3.3. The resistance between a pair of electrodes immersed in an infinite medium of conductivity σ is R. Show that the capacitance between the same pair of electrodes is $C = \varepsilon / R\sigma$ when the medium is changed to a lossless dielectric.

A general proof is required valid for any geometry.

3.4. Two lossy dielectric materials are joined together as shown in Fig. 3.18. Determine the voltages across each material and the surface charge density at the boundary if a voltage V is applied.

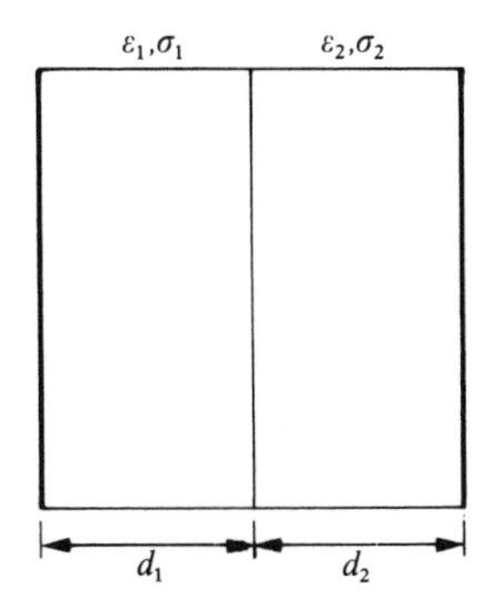

FIG. 3.18. Two lossy dielectric materials.

3.5. Solve eqn (3.47) under the condition that the electric field is zero at electrode 1 ($z = 0$).

3.6. Derive an expression for the magnetic field H at a point P distant a from the centre line of a long thin conducting strip of width b (Fig. 3.19) which carries a longitudinal current I uniformly distributed across its section.

3.7. (i) Prove that the magnetic field strength H at a point P distance R from the wire of finite length a (Fig. 3.20) is

$$H = \frac{I}{4\pi R}(\cos\alpha_1 - \cos\alpha_2).$$

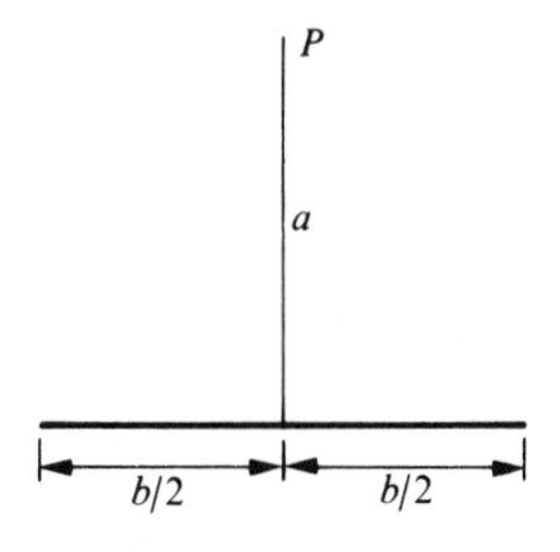

FIG. 3.19. A long conducting strip carrying a current I.

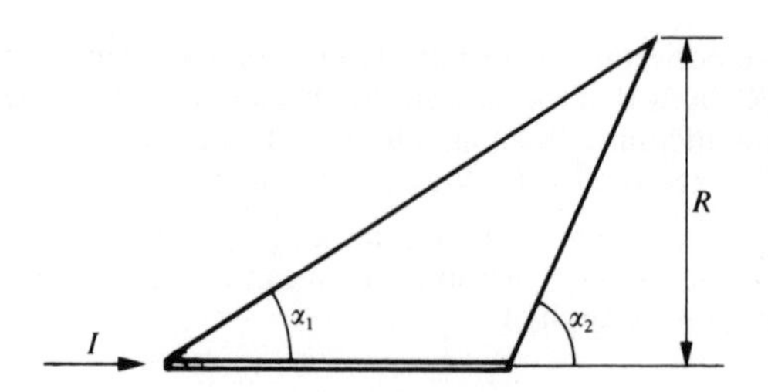

FIG. 3.20. A wire of finite length carrying a current I.

(ii) Deduce that the field strength at the centre of a square coil of side b, carrying a current I, is

$$H=\frac{2\surd(2)I}{b\pi}.$$

(iii) Determine the field strength at the centre of a current carrying loop which has the form of an n-sided regular polygon inscribed in a circle of radius a.

(iv) Show that the above result reduces to that of (ii) when $n=4$ and to eqn (3.65) (with $z=0$) when $n\to\infty$.

3.8. With the aid of the transformation

$$\frac{\varphi'-\varphi}{2}=\psi-\frac{\pi}{2}$$

bring eqn (3.55) to the form

$$A_\varphi=-\frac{\mu_0 Ia}{2\pi\{(R+a)^2+z^2\}^{\frac{1}{2}}}\left\{\frac{2}{k^2}\int_0^\pi(1-k^2\sin^2\psi)^{\frac{1}{2}}\,\mathrm{d}\psi+\left(1-\frac{2}{k^2}\right)\int_0^\pi\frac{\mathrm{d}\psi}{(1-k^2\sin^2\psi)^{\frac{1}{2}}}\right\},$$

where

$$k^2=\frac{4aR}{(R+a)^2+z^2}.$$

3.9. The complete elliptic integrals are defined as follows

$$K(k)=\int_0^{\pi/2}\frac{\mathrm{d}\psi}{(1-k^2\sin^2\psi)^{\frac{1}{2}}},\qquad E(k)=\int_0^{\pi/2}(1-k^2\sin^2\psi)^{\frac{1}{2}}\,\mathrm{d}\psi.$$

Express the vector potential of the previous example in terms of elliptic integrals and work out its value when $R=a=0{\cdot}1$ m, $I=1$ A, $z=a\surd 12$ (use Tables, e.g. Jahnke–Emde–Lösch).

3.10. When $k\ll 1$ the complete elliptic integrals may be approximated by

$$E(k)\cong\frac{\pi}{2}\left\{1-2\frac{k^2}{8}-3\left(\frac{k^2}{8}\right)^2\right\}$$

and

$$K(k)\cong\frac{\pi}{2}\left\{1+2\frac{k^2}{8}+9\left(\frac{k^2}{8}\right)^2\right\}.$$

Show that the vector potential obtained in Example 3.9 reduces to those of eqns (3.60) and (3.64) under the respective assumptions.

3.11. It is fairly simple to differentiate the elliptic integrals (see p. 49 of Jahnke–Emde–Lösch, for the formulae) so if you have the patience derive the magnetic field from the vector potential of Example 3.9 and determine its value for the data given there.

3.12. Eqn (3.70) gives the magnetic field on the axis of a solenoid of length l. Is the formula valid outside the solenoid?

3.13. The magnetization curve of a certain steel is given by

H	250	500	1000	2000 A m^{-1}
B	0·9	1·2	1·4	1·6 T.

An anchor ring of this material has a mean diameter of 255 mm. It is wound with 160 turns of wire carrying a current of 2·5 A. Calculate the flux density on the mean diameter. What current is required to produce the same flux density after a gap of 0·5 mm wide has been cut in the ring? What flux density is produced in the gap by a current of 7·5 A?

3.14. The electromagnet shown in section in Fig. 3.21 is designed to give radial magnetic flux density $\boldsymbol{B}$ in an annulus of radius a and width $\delta(\delta \ll a)$ when energized with constant voltage V. Its coil is wound from copper of conductivity σ and is located in the annular space inside the electromagnet which has the dimensions a, b, c shown in the figure. The copper can be assumed to be uniformly distributed across the section but it only occupies a fraction $\alpha(\alpha < 1)$ of the space available; the current density in the copper is to be J.

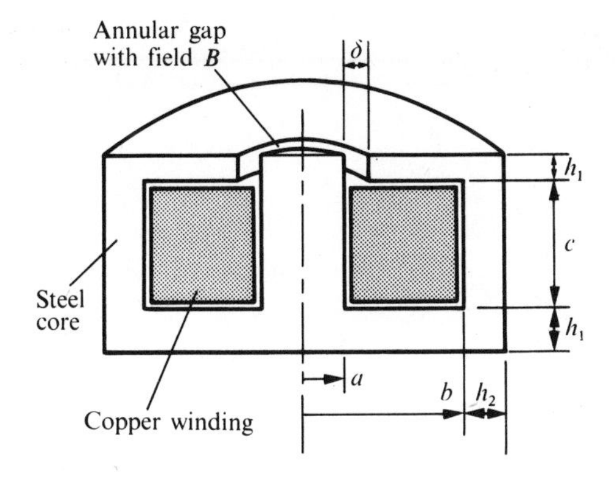

FIG. 3.21. The cross section of an electromagnet.

Neglecting flux leakage, and the contribution of the magnetic material to the line integral, calculate the dimension b, the size of the copper wire, and the number of turns in the winding when $B = 1$ T, $a = 20$ mm, $\delta = 1·5$ mm, $c = 0·1$ m, $V = 12$ V, $\sigma = 5·7 \times 10^7$ S m^{-1}, $\alpha = 0·65$, $J = 1·5 \times 10^6$ A m^{-2}. Explain why α must be less than unity and why the current density must be limited to J. Estimate values for the dimensions h_1, h_2 so that each section of the steel core shall be subject to approximately the same maximum magnetic flux density.

3.15. From eqn (3.90) and from the Biot–Savart law show that the force between two arbitrary current-carrying loops (Fig. 4.8, p. 100) is

$$\mathbf{F} = \frac{\mu_0 I_1 I_2}{4\pi} \oint\oint \frac{\mathbf{i}_{r_{12}}}{r_{12}^2} \, \mathbf{ds}_1 \, \mathbf{ds}_2,$$

where $\mathbf{i}_{r_{12}}$ is a unit vector in the direction between the current elements $\mathbf{ds}_1$ and $\mathbf{ds}_2$. (*Hint*; Use Stokes's theorem.)

4. Slowly varying phenomena

Whether Ampère's beautiful theory were adopted, or any other, or whatever reservation were mentally made, still it appeared very extraordinary, that as every electric current was accompanied by a corresponding intensity of magnetic action at right angles to the current, good conductors of electricity, when placed within the sphere of this action, should not have any current induced through them, or some sensible effect produced equivalent in force to such a current.

4. These considerations, with their consequence, the hope of obtaining electricity from ordinary magnetism, have stimulated me at various times to investigate experimentally the inductive effect of electric currents.

. . . it was ascertained, both at this and the former time, that the slight deflection of the needle occurring at the moment of completing the connexion, was always in one direction, and that the equally slight deflection produced when the contact was broken, was in the other direction;

as the particular action might be supposed to be exerted only at the moments of making and breaking contact, the induction was produced in another way. Several feet of copper wire were stretched in wide zigzag forms, representing the letter W, on one surface of a broad board; a second wire was stretched in precisely similar forms on a second board, so that when brought near the first, the wires should everywhere touch, except that a sheet of thick paper was interposed. One of these wires was connected with the galvanometer, and the other with a voltaic battery. The first wire was then moved towards the second, and as it approached, the needle was deflected. Being then removed, the needle was deflected in the opposite direction.

114. The relation which holds between the magnetic pole, the moving wire or metal, and the direction of the current evolved, i.e. *the law* which governs the evolution of electricity by magneto-electric induction, is very simple, although rather difficult to express.

MICHAEL FARADAY *Experimental reserches in electricity* 1831

Wir können hiebei eine mit den beschriebenen Einrichtungen in genauer Verbindung stehende grossartige und bisher in ihrer Art einzige Anlage nicht unerwähnt lassen, die wir unserm Herrn Prof. Weber verdanken. Dieser hatte bereits im vorigen Jahre von dem physikalischen Cabinet aus über die Häuser der Stadt hin bis zur Sternwarte eine doppelte Drahtverbindung geführt, welche gegenwärtig von der Sternwarte bis zum magnetischen Observatorium fortgesetzt ist. Dadurch bildet sich eine grosse galvanische Kette, worin der galvanische Strom, die an beiden Endpunkten befindlichen Multiplicatoren mitgerechnet, eine Drahtlänge von fast neuntausend Fuss zu durchlaufen hat.

Die Leichtigkeit und Sicherheit, womit man durch den Commutator die Richtung des Stroms und die davon abhängige Bewegung der Nadel beherrscht, hatte schon im vorigen Jahre Versuche einer Anwendung zu telegraphischen Signalisirungen veranlasst, die auch mit ganzen Wörtern und kleinen Phrasen auf das vollkommenste gelangen. Es leidet keinen Zweifel, dass es möglich sein würde, auf ähnliche Weise eine unmittelbare telegraphische Verbindung zwischen zweien eine beträchtliche Anzahl von Meilen von einander entfernten Örtern einzurichten: allein es kann natürlich hier nicht der Ort sein, Ideen über diesen Gegenstand weiter zu entwickeln.

CARL FRIEDRICH GAUSS *Magnetische Beobachtungen, Göttingische gelehrte Anzeigen* 1834 August 9

What led more more or less directly to the special theory of relativity was the conviction that the electromotive force acting on a body in motion in a magnetic field was nothing else but an electric field.

ALBERT EINSTEIN 1952 from a letter to the Michelson Commemorative Meeting of the Cleveland Physics Society.

4.1. The basic equations

IT is difficult to say what 'slowly' varying is until I give examples of 'fast' varying phenomena. Thus for a proper appreciation of the distinction between 'slow' and 'fast' you have to wait for electromagnetic waves to be introduced, discussed, and digested. For immediate use I offer only mathematics, but a little later (Section 4.3) I shall try to show the limitations imposed by the assumption of slow variation.

Mathematically, the definition is easy. As long as

$$\left|\frac{\partial \mathbf{D}}{\partial t}\right| \ll |\mathbf{J}|, \tag{4.1}$$

i.e. as long as the displacement current is negligible in comparison with the current of charged particles, we are in the 'slowly' varying region. Hence we disregard the displacement current term but have all the rest of eqns (1.1) to (1.7) as follows:

$$\nabla \times \mathbf{H} = \mathbf{J}, \tag{4.2}$$

$$\nabla \times \mathbf{E} = -\partial \mathbf{B}/\partial t, \tag{4.3}$$

$$\nabla \cdot \mathbf{D} = \rho, \tag{4.4}$$

$$\nabla \cdot \mathbf{B} = 0, \tag{4.5}$$

$$\mathbf{D} = \varepsilon \mathbf{E}, \tag{4.6}$$

$$\mathbf{B} = \mu \mathbf{H}, \tag{4.7}$$

$$\mathbf{F} = q(\mathbf{E} + \mathbf{v} \times \mathbf{B}). \tag{4.8}$$

There is only one equation we have not considered so far and that is eqn (4.3), which will probably look more familiar in another form. Integrating eqn (4.3) over a surface and applying Stokes's theorem to the left-hand side, we get

$$\oint \mathbf{E} \,.\, \mathrm{d}\mathbf{s} = -\frac{\partial}{\partial t}\int \mathbf{B} \,.\, \mathrm{d}\mathbf{S}, \tag{4.9}$$

where the line integral is over the closed contour of the chosen surface. Using the definition of magnetic flux (eqn (3.22)) the above equation may be written in the form

$$\oint \mathbf{E} \,.\, \mathrm{d}\mathbf{s} = -\frac{\partial \Phi}{\partial t}. \tag{4.10}$$

Remember eqn (2.5)

$$\mathbf{E} = -\nabla\phi, \tag{2.5}$$

the definition of the scalar potential in terms of the electric field. Substituting it into eqn (4.10) would always yield zero, and that is obviously incorrect because the right-hand side may be finite. Thus the definition of eqn (2.5) is no longer applicable or we should rather say it is no longer sufficient. We need something else besides the scalar potential for describing correctly a time-varying electric field. The additional term may be obtained by substituting the vector potential for **B** in eqn (4.3):

$$\nabla\times\mathbf{E} = -\frac{\partial}{\partial t}(\nabla\times\mathbf{A}) \tag{4.11}$$

and rearranging it in the form

$$\nabla\times\left(\mathbf{E}+\frac{\partial \mathbf{A}}{\partial t}\right) = 0. \tag{4.12}$$

For the above equation to be satisfied the expression in the bracket must be equal to zero apart from the gradient of a scalar function (remember $\nabla\times\nabla\phi = 0$). Thus the general form for the electric field is as follows

$$\mathbf{E} = -\frac{\partial \mathbf{A}}{\partial t} - \nabla\phi. \tag{4.13}$$

The gradient of a scalar potential is still there but we have in addition the time derivative of the vector potential.

When studying static electricity there was no need to make any distinction between voltage and potential difference. For the static case, by definition,

$$V = \phi_2 - \phi_1 = -\int_1^2 \mathbf{E} \,.\, \mathrm{d}\mathbf{s}. \tag{4.14}$$

For the time-varying case we still define voltage by the relation

$$V=-\int_1^2 \mathbf{E}\,.\,\mathrm{d}\mathbf{s}, \tag{4.15}$$

but now it takes the form

$$\begin{aligned} V &= \int \left(\nabla\phi+\frac{\partial \mathbf{A}}{\partial t}\right).\,\mathrm{d}\mathbf{s} \\ &= \phi_2-\phi_1+\frac{\partial}{\partial t}\int_1^2 \mathbf{A}\,.\,\mathrm{d}\mathbf{s}. \end{aligned} \tag{4.16}$$

There is no reason why the line integral of the vector potential between two arbitrary points should be independent of the path. This must be kept in mind when considering time-varying fields. It is no longer unambiguous to talk about the voltage between two points. It will, in general, depend on the path chosen.

For the static case the line integral of the electric field disappears when taken over a closed path. For the time-varying case $\oint \mathbf{E}\,.\,\mathrm{d}\mathbf{s}$ is, in general, different from zero. We shall introduce now the notation

$$\mathscr{E}=\oint \mathbf{E}\,.\,\mathrm{d}\mathbf{s}, \tag{4.17}$$

where $\mathscr{E}$ is the so-called *electromotive force*. It is rather unfortunate to call it a force because it isn't one. At the same time one can have sympathy with those who devised the term because it is the tangential force per unit charge integrated over a closed path. In other words it is the work done by taking a unit charge round a circuit. The definition, eqn (4.17), though often used, is regrettably not sufficiently general. When the closed path is in a wire loop and the loop is in motion then a force due to the magnetic field is present as well, giving rise to a finite amount of work in the same manner. The more general definition will be discussed in Section 4.7.

For the time being we shall use eqn (4.17), which substituted into eqn (4.10) will yield

$$\mathscr{E}=-\frac{\partial \Phi}{\partial t}, \tag{4.18}$$

a restricted form of Faraday's law (for its general form, see Section 4.7). This is valid for the case when the flux linking a stationary circuit varies as a function of time.

The dimension of $\mathscr{E}$ is that of voltage, so it is not surprising that many people refer to it as voltage or induced voltage, a usage into which I often lapse myself. However, when we want to emphasize the ability of $\mathscr{E}$ to put charges into motion, it seems preferable to accord to it its full title, or at least its popular abbreviation in the form of e.m.f.

We have now derived a new law for slowly varying phenomena. What can we say about the laws derived in the last chapter? Are Ampere's law and Biot–Savart's law still valid? And the relationship derived between the current density and the vector potential (eqn (3.15)), is that still valid? Yes, all of them are still true as a good approximation as long as the inequality (4.1) stands. A slowly varying current will produce a vector potential (or a magnetic field) varying at the same rate.

4.2. The electric field due to a varying magnetic field

Let us consider a two-dimensional case where the magnetic field at a given moment is constant within a cylinder of radius a, and is zero outside this cylinder. We shall further assume a sinusoidal time variation so that

$$\begin{aligned} \mathbf{B} &= B_0 \sin \omega t \mathbf{i}_z \qquad R \leqslant a, \\ \mathbf{B} &= 0 \qquad\qquad\quad\;\; R > a. \end{aligned} \tag{4.19}$$

The assumed magnetic field is independent both of z and of the azimuth angle φ, hence the electric field follows the same pattern; it is independent of z and φ and depends on R only. Then eqn (4.3) yields the scalar differential equations

$$\frac{1}{R}\frac{\partial}{\partial R}(RE_\varphi) = -B_0\omega \cos \omega t \qquad R \leqslant a, \tag{4.20}$$

$$\frac{1}{R}\frac{\partial}{\partial R}(RE_\varphi) = 0 \qquad R > a. \tag{4.21}$$

We find by inspection that the solutions are

$$E_\varphi = -\frac{B_0\omega}{2} R \cos \omega t \qquad R \leqslant a, \tag{4.22}$$

$$E_\varphi = \frac{\text{constant}}{R} \qquad R > a. \tag{4.23}$$

The constant may be determined from the continuity of the electric field at $R = a$ yielding

$$E_\varphi = -\frac{B_0 \omega a^2}{2R} \cos \omega t \qquad R > a. \tag{4.24}$$

This is quite interesting. A time-varying magnetic field creates an electric field as suggested by eqn (4.3), but this is not necessarily a local relationship. The magnetic field may be confined to a certain part of space but the resulting electric field will pervade *all* space.

What can we say about the line integral of the electric field? It follows from eqn (4.9) that it is finite if the path encloses the time-varying flux but zero otherwise. How does this appear in practice? What happens if we place a conducting wire into the electric field? The mobile charges in the wire are not the least concerned about the origin of the electric field. They react in the same way whether the electric field is due to static charges or to a time-varying magnetic field; under the force $q\mathbf{E}$ they rearrange themselves so as to cancel the electric field inside the conducting material as shown in Fig. 4.1(a). There will now be an additional electric field (say $\mathbf{E}_c$) due to the presence of these charges. Note however that

$$\oint \mathbf{E}_c \cdot \mathrm{d}\mathbf{s} = 0, \tag{4.25}$$

hence the charges do *not* interfere with the line integral of the electric field around a closed path. It is still true that

$$\oint \mathbf{E} \cdot \mathrm{d}\mathbf{s} = -\frac{\partial \Phi}{\partial t}, \tag{4.26}$$

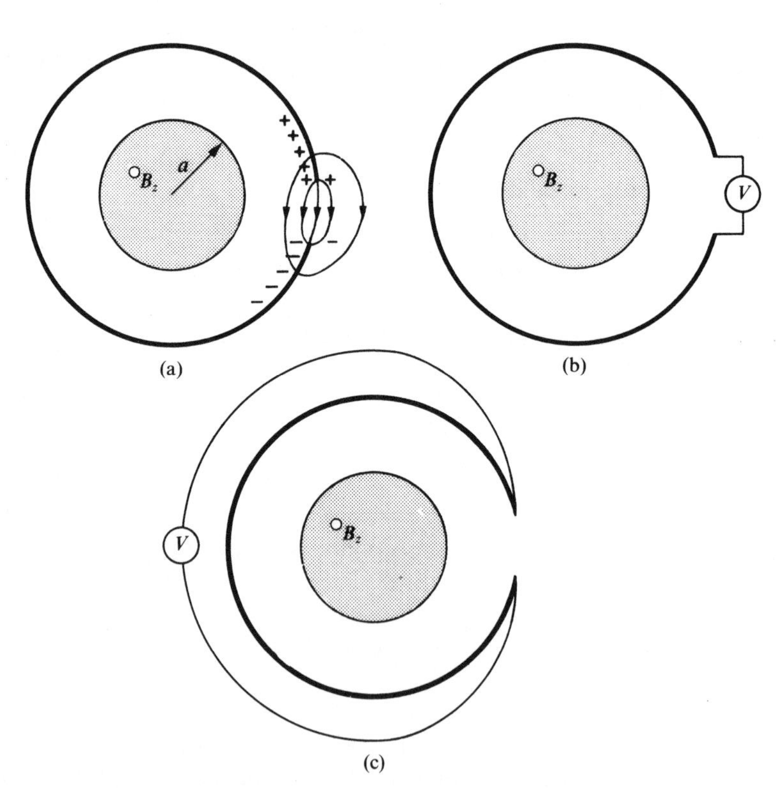

FIG. 4.1. A wire loop surrounding a time-varying magnetic field inside the shaded cylinder of radius a. (a) The charge distribution and the corresponding electric field at a given time. (b) Finite voltage is measured when the loop encloses the magnetic field. (c) Zero voltage is measured when the loop does not enclose the magnetic field.

whether we regard **E** as the original field (without the contribution of the charges) or as if $\mathbf{E}_c$ were added to the original field.

So what is the voltage we are going to measure? Let us take an ideal voltmeter (one that draws no current) and measure the voltage across the terminals of the wire. The voltage measured still depends on the path. We measure finite voltage if the time-varying flux is enclosed (Fig. 4.1(b)) and zero voltage if no flux is enclosed (Fig. 4.1(c)).

What happens if we have N loops (Fig. 4.2(a)) round the varying flux? Surely, we will have an induced voltage $-\partial\Phi/\partial t$ in each one of them. If we

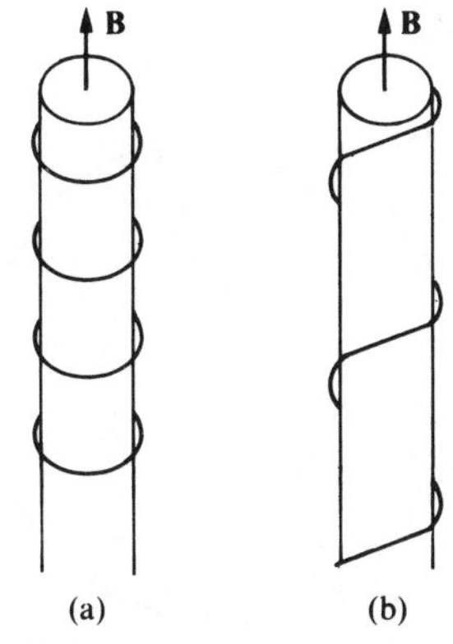

FIG. 4.2. (a) N turns and (b) helical winding around the cylindrical magnetic field.

connect the loops then the total induced voltage will be the algebraic sum of the individual voltages. If we make up a helical coil (Fig. 4.2(b)) the wires are going round always in the same direction so the voltages simply add and we may rewrite eqn (4.13) in the form

$$V = -N\frac{\partial\Phi}{\partial t}, \tag{4.27}$$

where Φ is the flux enclosed by each turn.

Let us go now one step further and consider a resistive wire ring (Fig. 4.3(a)). The force on the mobile charges is still $q\mathbf{E}$ but now the charges may follow the electric field all the way around the ring. A current is set up corresponding to the $\mathbf{J} = \sigma\mathbf{E}$ relationship. Then

$$V = \int \mathbf{E}\,.\,\mathrm{d}\mathbf{s} = \frac{J}{\sigma}2\pi b = IR_l, \tag{4.28}$$

where R_l is the resistance of the loop. As may be expected the ohmic voltage drop in the wire will be equal to the induced voltage.†

† This is under the assumption that the magnetic field produced by the current in the ring is negligible. The general case will be treated later in Section 4.3.

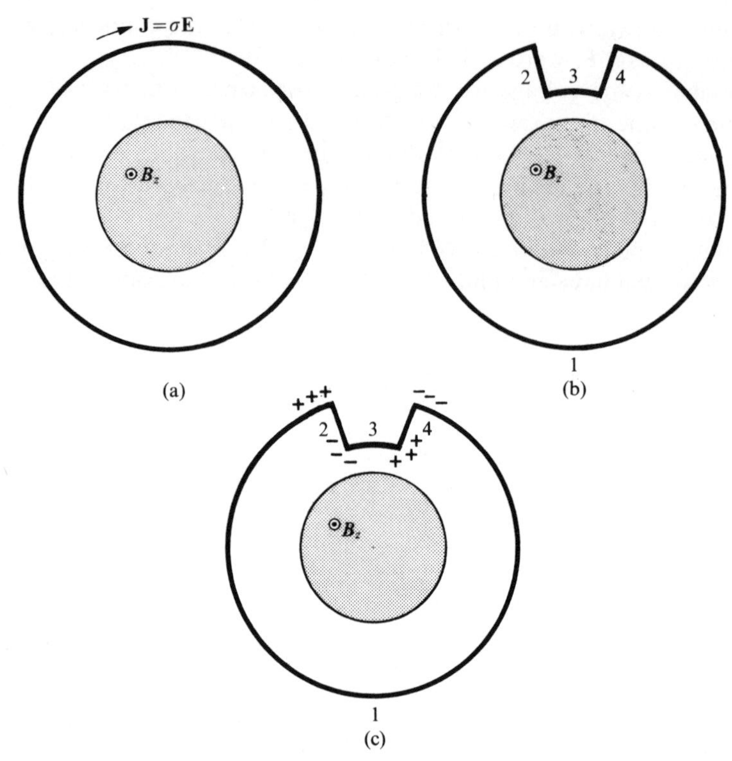

FIG. 4.3. (a) The current flowing in the resistive wire is determined by the induced electric field. (b) The induced electric field is different in the four sections shown. (c) Charges arise so as to make the tangential component of the electric field (and hence the current density) equal at every point of the loop.

Let us next evaluate the line integral of the electric field over the closed mathematical curve shown in Fig. 4.3(b). As long as the total time-varying magnetic flux is enclosed, the line integral $\oint \mathbf{E} \cdot d\mathbf{s}$ will remain the same as in the previous example. Note however that the electric field is different on Sections 1 and 3 and the line integrals on Sections 2 and 4 vanish altogether (because the path is perpendicular to the electric field).

Let us replace now the mathematical curve by a thin wire. What will determine the current flowing in the wire? The tangential component of the electric field along the wire. But then we obtain the answer that the current is larger in Section 3 than in Section 1, and no current flows in Sections 2 and 4. Is that possible? According to eqn (4.2) that is *not* possible. Taking the divergence of both sides of eqn (4.2) we get

$$\nabla \cdot \mathbf{J} = 0, \tag{4.29}$$

which may be recognized as the continuity equation for an incompressible fluid.† It means in our case (the same that commonsense would suggest) that the current in the loop must everywhere be the same. It varies of course as a function of time but not as a function of the spatial coordinates. So **J** must remain constant, but it can only remain constant if the tangential component of the electric field is constant as well. Hence charges must appear (see Fig. 4.3(c)), producing an electric field $\mathbf{E}_c$ which will counteract the induced electric field $\mathbf{E}_i$ so that the resultant electric field $\mathbf{E}=\mathbf{E}_i+\mathbf{E}_c$ is constant everywhere along the wire.

What difference will it make if the wire loop is placed in the magnetic field as shown in Fig. 4.4? There will be a current produced by the electric field as

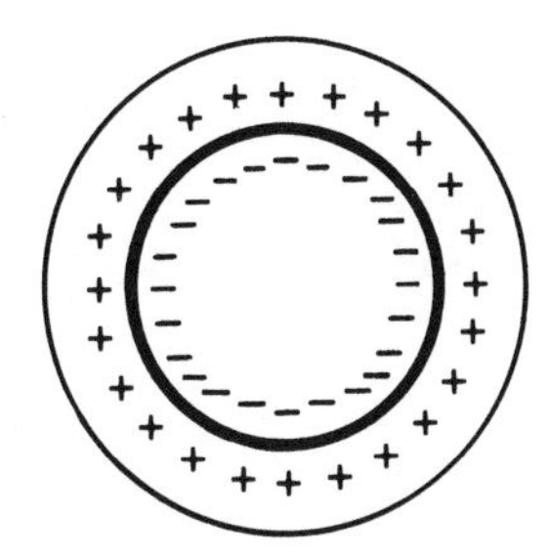

FIG. 4.4. Radial displacement of charge in a loop placed into the magnetic field.

before; the new feature is the appearance of a $q\mathbf{v}\times\mathbf{B}$ force on each charge element. Since the velocity of the charge carriers is in the azimuthal direction, the force will act radially. It is the same story again. The charges cannot leave the wire so they will accumulate at the outer and inner surface of the ring until a radial electric field is produced that will cancel the force due to the magnetic field.

In conclusion, I wish to emphasize that all the charge rearrangements discussed in this section occur very fast, much faster than the period of oscillation of the magnetic field.

4.3. Inductance and mutual inductance

We shall start again with a resistive ring but make two modifications. The externally impressed voltage need not come from a time-varying magnetic field as in the previous section, it may be produced by a signal generator. Secondly, we shall take into account the voltage induced by the current flowing in the ring. If you solve the field problem‡ and exercise due care in getting the signs right, you will find that the induced voltage is such as to oppose the voltage that created it. For this reason it is often called a *back e.m.f.*

† A more detailed discussion of the continuity equation will come in Section 5.4.
‡ Alternatively, you may appeal to Lenz's law.

The resulting equation is

$$V_{\text{ext}} = RI - V_{\text{i}}. \tag{4.30}$$

But according to Faraday's law

$$V_{\text{i}} = -\frac{\partial \Phi}{\partial t}, \tag{4.31}$$

hence

$$V_{\text{ext}} = RI + \frac{\partial \Phi}{\partial t}. \tag{4.32}$$

We shall now define *self-inductance* by the relationship

$$LI = \Phi. \tag{4.33}$$

Substituting the above definition into eqn (4.32) and changing from $\partial/\partial t$ to $\mathrm{d}/\mathrm{d}t$ we get†

$$V_{\text{ext}} = RI + L\frac{\mathrm{d}I}{\mathrm{d}t}, \tag{4.34}$$

as known from circuit theory. Eqn (4.34) is suitable for the measurement of L. We impose a sinusoidal V_{ext}, we know R (from calculations or measurement), we measure I, calculate $\mathrm{d}I/\mathrm{d}t$, and then L, being the only unknown, is determined. Can we calculate L? Not easily. Even for such a simple geometry as the ring there are lots of difficulties. In Section 3.7 we managed to get simple expressions for the magnetic field far away from the ring and in the vicinity of the axis. But in order to calculate the flux across the ring we need to know either the magnetic field at every point in the interior of the ring or the vector potential along the perimeter of the ring. Eqn (3.56) is complicated enough and that does not even take account of the finite diameter of the wire. So the problem is pretty complicated. Reference books give the answer

$$L = \mu a\left(\ln\frac{8a}{d} - 2\right), \tag{4.35}$$

where d is the diameter of the wire. This looks simple enough but, believe me, a lot of sweat has gone into producing that formula.

Eqn (4.34) is of course true for any closed circuit upon which a time-varying voltage is impressed. In general, R is the resistance of the circuit and L may be obtained from eqn (4.33), where Φ is the flux enclosed by the circuit.

† There are no sophisticated reasons necessitating this change, we are not going over from a stationary to a co-moving frame of reference, we only wish to reproduce the notations of circuit theory where, time being the only variable, $\mathrm{d}/\mathrm{d}t$ is used.

Can one determine the inductance of any of the practical configurations with relative ease? Yes, there are a few, e.g. a two-wire transmission line which is nothing else but two parallel wires carrying opposite currents. For two infinite line currents the magnetic field was given by eqn (3.53). We get the flux per unit length by integrating the flux density over the space between the wires (assumed to be infinitely thin), as follows:

$$\Phi = \mu_0 \int_0^b H_\varphi \, \mathrm{d}x = \frac{\mu_0 I}{2\pi} \int_0^b \left(\frac{1}{x} + \frac{1}{b-x} \right) \mathrm{d}x. \tag{4.36}$$

Unfortunately, the above integral diverges. What's wrong? Obviously the limits: we should not have taken the wires infinitely thin. These infinities are our best friends and worst enemies. By taking things infinitely long and infinitely thin and infinitely something else, we can arrive at simple formulae. When later we want to use those formulae it turns out not infrequently that we have some divergent results. In the present case the remedy is clear and easy. We need to assume a finite wire diameter, and we can do that without landing in a sea of further complications. This is because wires used in practice (e.g. copper) are non-magnetic: we don't need to worry about boundary conditions at all. The only difference between copper wire and air is that the former carries the current. There is no reason now for the current density to deviate from uniform distribution over the cross-section, so eqn (3.53) still correctly describes the magnetic field between the wires. Consequently, the flux per unit length between the wires is

$$\Phi = \frac{\mu_0 I}{2\pi} \int_a^{b-a} \left(\frac{1}{x} + \frac{1}{d-x} \right) \mathrm{d}x = \frac{\mu_0 I}{\pi} \ln \frac{b-a}{a}, \tag{4.37}$$

where a is the radius of the wire. Is this the total flux? No, because the magnetic field can penetrate the wires. The contribution of a single wire may be determined with reference to Fig. 4.5. At a radius r the amount of current enclosed is $(r/a)^2 I$, hence the magnetic field from Ampere's law is

$$H = \frac{1}{2\pi r} \left(\frac{r}{a} \right)^2 I, \tag{4.38}$$

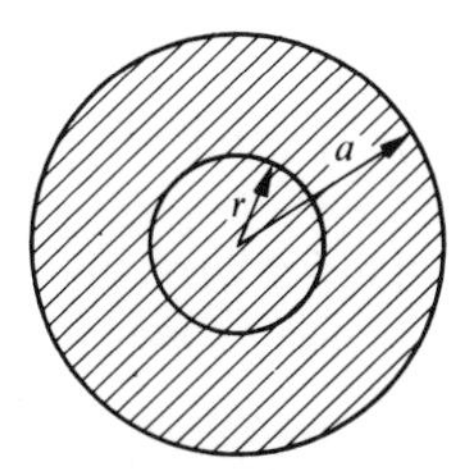

FIG. 4.5. The current carried within a radius r is $(r/a)^2 I$.

yielding for the flux per unit length inside the material

$$\Phi_{\text{int}} = \frac{\mu_0 I}{2\pi a^2}\int_0^a r\,\mathrm{d}r = \frac{\mu_0 I}{4\pi}. \tag{4.39}$$

If $b \gg a$ then this internal flux may be neglected giving the often quoted formula† for the inductance per unit length of a two-wire transmission line

$$L = \frac{\mu_0}{\pi}\ln\frac{b}{a}, \tag{4.40}$$

but remember there is an internal flux as well (and a corresponding internal inductance), which may not always be negligible.

Next we shall look at a long solenoid filled with a high-permeability magnetic material as shown in Fig. 3.9 (p. 73). Then the magnetic field inside the material (from eqn (3.72)) is

$$H = \frac{NI}{l}, \tag{4.41}$$

uniform across the cross-section (of area S_0) of the core. Hence the magnetic flux produced inside the solenoid is

$$\Phi = BS_0 = \frac{\mu NIS_0}{l}. \tag{4.42}$$

The voltage induced in *each* turn of the solenoid is $\mathrm{d}\Phi/\mathrm{d}t$ hence the self-inductance comes to

$$L = \frac{N\Phi}{I} = \mu\frac{N^2S_0}{l}. \tag{4.43}$$

Let us tap now our solenoid at a certain point (Fig. 4.6(a)) so that

$$N = N_1 + N_2 \quad \text{and} \quad \frac{N_1}{l_1} = \frac{N_2}{l_2} = \frac{N}{l}. \tag{4.44}$$

We have now two separate solenoids with N_1 and N_2 turns respectively. How could we determine their inductances? Nothing can be simpler, just use eqn (4.43) above and get

$$L_1 = \mu\frac{N_1^2S_0}{l_1} = \frac{N_1}{N}L \quad \text{and} \quad L_2 = \mu\frac{N_2^2S_0}{l_2} = \frac{N_2}{N}L. \tag{4.45}$$

I am afraid the above formulae are wrong because our new solenoids don't look the same as the old one. As shown in Figs 4.6(b) and (c) the new

† Another reason for the validity of this formula is that for frequencies at which a two-wire transmission line is used the currents are confined near to the surface of the wire (owing to the so-called *skin effect* to be discussed later, in Section 5.11) reducing further the significance of the internal flux.

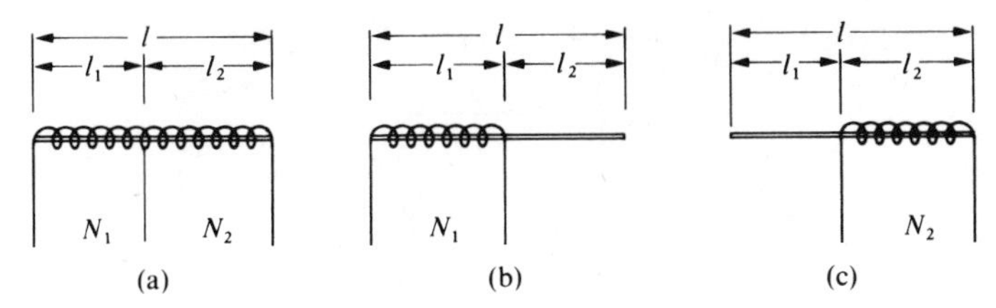

FIG. 4.6. Solenoids wound on the same iron core.

solenoids have a core of length l, although the wiring is confined to lengths l_1 and l_2 respectively. Since the magnetic material is still endowed with the property of concentrating the field lines in itself, the flux stays constant for the whole length of the core, yielding

$$L_1 = \mu_0 \frac{N_1^2 S_0}{l} = \left(\frac{N_1}{N_0}\right)^2 L \quad \text{and} \quad L_2 = \mu_0 \frac{N_2^2 S_0}{l} = \left(\frac{N_2}{N}\right)^2 L. \tag{4.46}$$

If the core makes up a closed circuit as shown in Fig. 4.7(a) then l should be taken as the length along the dotted line and eqn (4.43) is still applicable.

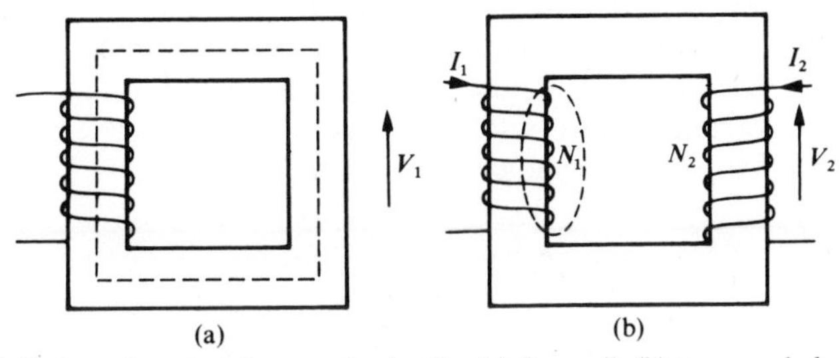

FIG. 4.7. Coils wound on closed magnetic circuits. (a) One coil; (b) two coupled coils (dotted lines show a magnetic field line that crosses coil 1 but not coil 2).

In all of our examples so far we could calculate the flux through a simple surface, but how should we take the surface when there is no magnetic material to guide the field lines and the wire itself is of a complicated shape? Then even the choice of the surface (having the wire as its boundary) might tax to the limit one's meagre imagination, let alone the calculation of the flux. The practical answer is 'don't calculate the inductance, measure it', but if for some reason you must do the calculation, the best approach may be to abandon the definition of eqn (4.33) in favour of the equivalent energetic definition (see Section 4.12)

$$\tfrac{1}{2}LI^2 = \text{total magnetic energy}. \tag{4.47}$$

It might be simpler to find the total energy in space than the flux crossing a given surface.

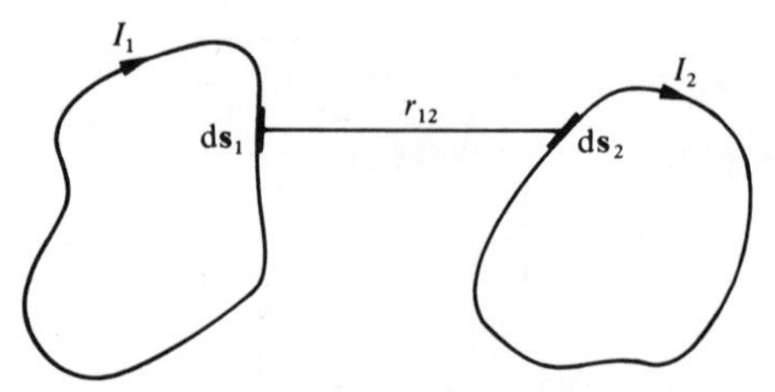

FIG. 4.8. Two coupled loops.

We shall now go over to the definition of mutual inductance with the aid of Fig. 4.8. The current I_1 flowing in loop 1 will produce a magnetic flux Φ_{12} through loop 2 leading to the definition

$$M_{12}=\frac{\Phi_{12}}{I_1}. \tag{4.48}$$

There is a similar definition for M_{21} and it can be shown (see Example 4.4) that $M_{12}=M_{21}=M$.

If I_1 is time-varying, the voltage induced in loop 2 is $M(\mathrm{d}I_1/\mathrm{d}t)$. The sign of M depends on the convention adopted for the direction of currents in the two loops. Circuit engineers take the mutual inductance invariably positive and denote the direction of the induced voltage by a dot in the circuit diagram. A current flowing into the inductance L_1 at the dot produces a voltage $V_2=M(\mathrm{d}I_1/\mathrm{d}t)$ as shown in Fig. 4.9.

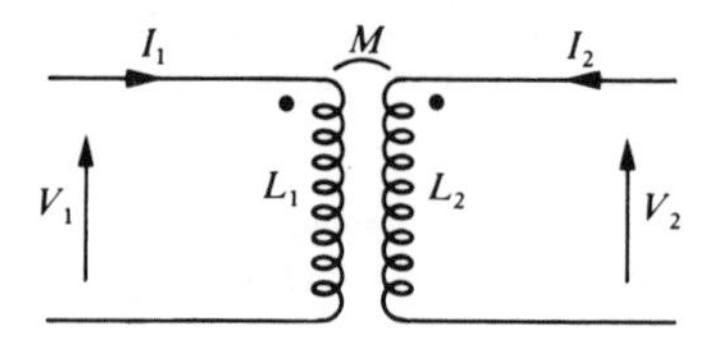

FIG. 4.9. The circuit notation for two coupled inductors.

Let us determine now the mutual inductance between two coils wound on the same magnetic core (Fig. 4.7(b)). The flux produced by a current in coil 1 is given by eqn (4.42),

$$\Phi=\frac{\mu N_1 I_1 S_0}{l}, \tag{4.49}$$

where the subscript 1 referring to coil 1 has been attached to the current and number of turns.

Assuming now that all the flux produced by coil 1 will pass through coil 2 as well, we get for the voltage in coil 2

$$V_2=M\frac{\mathrm{d}I_1}{\mathrm{d}t}=N_2\frac{\mathrm{d}\Phi}{\mathrm{d}t}=\frac{\mu N_1 N_2 S_0}{l}\cdot\frac{\mathrm{d}I_1}{\mathrm{d}t}, \tag{4.50}$$

whence

$$M = \frac{\mu N_1 N_2 S_0}{l}. \tag{4.51}$$

It is interesting to note that

$$M^2 = L_1 L_2. \tag{4.52}$$

If some of the flux produced by coil 1 does not pass through coil 2 (see Fig. 4.7(b)) then Φ appearing in eqn (4.50) is smaller leading to a smaller value of mutual inductance. Thus when some flux 'leaks away' eqn (4.52) becomes an inequality:

$$M^2 < L_1 L_2. \tag{4.53}$$

Alternatively, one may introduce a *coupling coefficient* $k \leq 1$ and rewrite eqns (4.52) and (4.53) in the form

$$M = k\sqrt{(L_1 L_2)}. \tag{4.54}$$

The above relationship turns out to be true for any two magnetically coupled circuits. The proof will be provided in Section 4.12.

Next we shall determine the mutual inductance between two concentric rings at a distance h apart (Fig. 3.17, p. 82) under the assumption that the upper ring is small. We need to find the magnetic flux through ring 2 due to the current in ring 1. We can use the formulae derived in Section 3.7 for the vector potential and magnetic field in the vicinity of the axis. With the aid of eqns (3.64) and (3.65) we obtain the magnetic flux in the form

$$\Phi_{12} = a_2 \int A_\Phi \mathrm{d}\varphi = \frac{\pi}{2} \frac{\mu_0 I_1 a_1^2 a_2^2}{(a_1^2 + h^2)^{\frac{3}{2}}} \tag{4.55}$$

from the vector potential, and in the form

$$\Phi_{12} = \mu_0 \int H_z \,\mathrm{d}S = \mu_0 H_z a_2^2 \pi = \frac{\pi}{2} \frac{\mu_0 I_1 a_1^2 a_2^2}{(a_1^2 + h^2)^{\frac{3}{2}}} \tag{4.56}$$

from the magnetic field, giving of course identical results. The mutual inductance is then

$$M_{12} = \frac{\Phi_{12}}{I_1} = \frac{\mu_0 \pi}{2} \frac{a_1^2 a_2^2}{(a_1^2 + h^2)^{\frac{3}{2}}}. \tag{4.57}$$

Note that we have not so far specified the wire thickness of either ring. For the mutual inductance it is not needed.

Finally, I want to say a few words about the assumption of 'slowly varying currents' in connection with the last example. As you know, any electromagnetic disturbance (any change in anything) propagates with the velocity of light, something we have so far neglected to take into account.

When calculating the mutual inductance we assumed that the magnetic field due to I_1 appears *instantaneously* at the second ring. In fact it takes time, the maximum time being equal to $\tau = l/c$, where l is the largest possible distance between any two points in Fig. 3.17. If the current I_1 may be considered constant during the time τ then we are entitled to talk about slow variation.

For sinusoidal current variation the condition is that the period of oscillation T should be much larger than τ. Using the relationships $T = 1/f$ and $c = f\lambda$ (where f and λ are the frequency and wavelength of oscillations) the condition may be written in the alternative form

$$l \ll \lambda. \tag{4.58}$$

This is clear. As long as the maximum dimension involved is small in comparison with the wavelength, we are in the slowly varying region.

4.4. Kinetic inductance

We have defined inductance by eqn (4.33) without discussing its effect upon the current–voltage relationship. Let me now briefly recall what an inductance does in a circuit. It delays things. If we apply a voltage the current will appear after some delay. If we switch off the voltage the current will disappear after some delay.

Is there anything else in an electrical circuit that behaves that way? That reminds me meeting an American friend of mine after he completed his grand tour of Europe. 'D'you know', he told me, 'that I can ask for the check in seventeen languages?' 'Eighteen', I said, 'I bet you forgot to include "bill".' As the above story shows it is often difficult to think of the obvious. In order to have a current, charge carriers must be accelerated, and it takes time to accelerate particles of finite mass. Hence the current will necessarily lag behind the voltage causing its rise.

Let us put now the relations in mathematical form. We shall write up Newton's equation for the case when the force is provided by an electric field and friction is present,

$$m\left(\frac{dv}{dt} + kv\right) = eE, \tag{4.59}$$

where k is a contant characterizing friction (to be related presently to more familiar constants).

For a cylindrical piece of conducting material of length l and cross-section S_0

$$V = El \quad \text{and} \quad I = S_0 \rho v, \tag{4.60}$$

which substituted into eqn (4.59) leads to

$$V = \frac{lm}{S_0 e \rho}\left(kI + \frac{dI}{dt}\right). \tag{4.61}$$

We may formally write

$$V = R_\kappa I + L_\kappa \frac{dI}{dt}, \tag{4.62}$$

where

$$R_\kappa = \frac{lmk}{S_0 e\rho} \tag{4.63}$$

is the kinetic resistance, and

$$L_\kappa = \frac{lm}{S_0 e\rho} \tag{4.64}$$

is the kinetic inductance.

If you think a little about it you will be able to convince yourself that R_κ is nothing other than the old familiar resistance (proportional to length; inversely proportional to cross-section) derived in a different manner. Whether one defines a conductance or introduces a friction term they are just two different ways of expressing the empirical fact that the electrons' velocity does not go on increasing indefinitely in response to a driving electric field.

Does the same apply to the second term? Is that also a different derivation of the inductance? No, in deriving eqn (4.61) we have not talked about magnetic fields and induction at all. The kinetic inductance owes its existence to inertia.

Why is eqn (4.64) so little known? Because under normal circumstances this kinetic inductance is negligible. Comparing eqns (4.63) and (4.64) you may see that $L_\kappa = R_\kappa / k$, and if you determine k from the identity

$$R_\kappa = R, \qquad \text{i.e.} \ \frac{lmk}{S_0 e\rho} = \frac{l}{S_0 \sigma} \tag{4.65}$$

you will find for copper that $1/k \approx 10^{-14}$ s. So you can see that the time constant involved is very small. The kinetic inductance does, however, acquire importance in superconductors, where the resistance disappears altogether. It may be seen by comparing eqns (4.64) and (4.35) that for a ring made of sufficiently thin wire the kinetic inductance may exceed the ordinary magnetic inductance. And this may very well occur in practice because superconductors used in integrated circuits can have cross-sections less than 10^{-12} m^2. It may be said in general that the designer of a superconducting circuit needs to worry about the response time and energy of superconducting electrons.

4.5. The transformer

Let us return to the configuration of two coils on a magnetic core (Fig. 4.7) and apply a sinusoidal voltage to coil 1 from an ideal voltage generator (having zero internal resistance). In response to the applied voltage there will be a current producing a flux that will induce voltages both in coils 1 and 2. If the coils are lossless then the voltage induced in coil 1 must balance the applied voltage, yielding

$$V_1 = N_1 \frac{d\Phi}{dt}. \tag{4.66}$$

Assuming for the moment that all the flux is contained within the magnetic core (so that the amount of flux crossing coils 1 and 2 is the same), the open circuit voltage in coil 2 is

$$V_2 = N_2 \frac{d\Phi}{dt}, \tag{4.67}$$

leading to the familiar relationship

$$\frac{V_1}{V_2} = \frac{N_1}{N_2} = n. \tag{4.68}$$

Under open-circuit conditions there is no current flowing in coil 2. The current in coil 1 may be obtained from the relationship

$$V_1 = L_1 \frac{dI_m}{dt}. \tag{4.69}$$

For an applied voltage of

$$V_1 = V_{10} \cos \omega t \tag{4.70}$$

we get

$$I_m = \frac{V_{10}}{\omega L_1} \sin \omega t. \tag{4.71}$$

What happens if we connect a resistance R_L across the terminals of coil 2? A current I_2 will flow. But the total flux should not be affected because it is still related to the applied voltage by eqn (4.66). Hence an additional current I_1' will be drawn from the generator so as to satisfy the equation

$$N_1 I_1' + N_2 I_2 = 0. \tag{4.72}$$

If L_1 is sufficiently large so that $I_m \ll I_1'$ then I_1' may be taken as the total primary current I_1, leading to the relationship

$$\frac{I_1}{I_2} = -\frac{N_2}{N_1} = -\frac{1}{n}. \tag{4.73}$$

Eqns (4.68) and (4.73) are known as the relationships valid for *ideal transformers.* They are certainly very useful for an engineer because they relate practical requirements (e.g. how large the voltage should be at the secondary) to design parameters (what the turns ratio should be).

How are transformers represented by circuit engineers? Well, the circuit representation shown in Fig. 4.9 is perfectly adequate for lossless transformers, but it would not yield the equations for an ideal transformer, not even for the case of perfect coupling, $k = 1$. For an ideal transformer we need to introduce a separate notation which we shall choose in the form shown in Fig. 4.10. The relationship between the two kinds of notations is not an

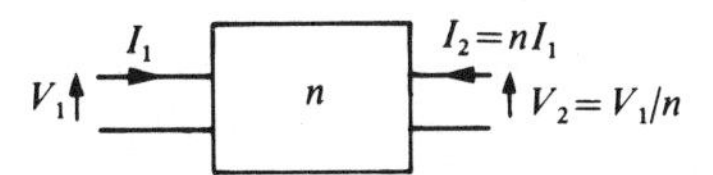

FIG. 4.10. A notation for an ideal transformer.

obvious one. It may be shown (using the criterion that they lead to the same set of equations) that by adding the so-called *leakage inductances* to the ideal transformer the two representations (Fig. 4.11) become equivalent.

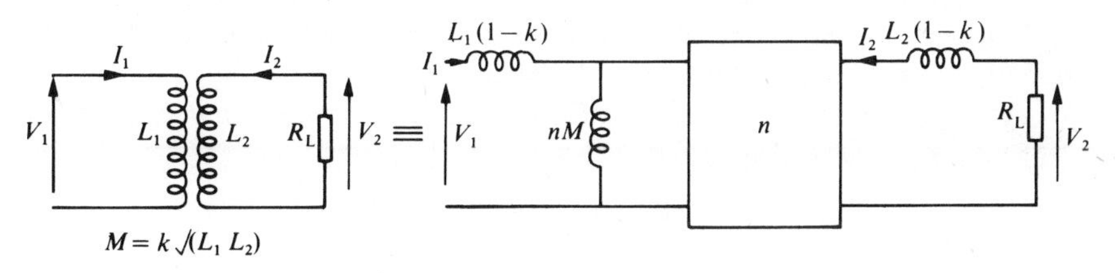

FIG. 4.11. Two equivalent representations of a transformer.

One may add copper losses (resistive losses in the wires) in a similar manner. If you want to include iron losses (occurring in the magnetic core due to the periodically changing flux) as well, you should better consult a book having a bigger section on transformers.

4.6. Relative motion of conducting wire and magnetic field

Up to now all the wires have been stationary, and the magnetic field has varied as a function of time. Let us see now what happens when the wires are made to move with uniform velocity. As our first example we shall take a straight piece of wire moving perpendicularly to the direction of a static magnetic field. As a consequence there will be a force $q\mathbf{v} \times \mathbf{B}$ acting on the charges in the wire; the mobile charges will be displaced until the arising electric field produces an equal and opposite force.

Let us investigate now the converse problem when the wire is stationary but the equipment producing the magnetic field is moved lock, stock, and barrel with the same velocity in the opposite direction. One's first impression is (or should be) that only relative motion counts, hence the force upon the charges should be the same. Formally, we would have the same force if we assumed (as many textbooks do) that a magnetic field moving with a velocity **u** gives rise to a force

$$\mathbf{F} = -q(\mathbf{u} \times \mathbf{B}). \tag{4.74}$$

This is a way out of the problem but not a way open to us. We have started by writing down Maxwell's equations and claimed that they will provide the solution to any electromagnetic problem. We are not entitled to introduce a new force. According to our equations a magnetic field, whether it moves or not, cannot produce a force on a stationary charge.

What should we do? Well, perhaps the trouble is that we wrote down Maxwell's equations in a stationary frame of reference. In order to tackle the present problem we should change to a coordinate system moving with the magnetic field. This is certainly a possible approach, and one we shall adopt towards the end of the course when discussing relativity, but such a transformation (though convenient) should not be necessary. Maxwell's equations written in the stationary frame of reference should be perfectly capable of dealing with the problem. Why is there a doubt at all? Some doubts may arise by considering the following problem.

Let us postulate the existence of a static magnetic field that is constant over a finite region of space, say

$$\begin{aligned} \mathbf{B} &= B\mathbf{i}_z \qquad |x| < a, \\ \mathbf{B} &= 0 \qquad\quad |x| > a, \end{aligned} \tag{4.75}$$

and assume that this magnetic field is bodily moved in the $+x$ direction with a velocity **u**. What will happen to the charges in the stationary wire situated at (say) $x = 0$? One may argue that by moving the magnetic field nothing has changed at the position of the wire. The magnetic field is unchanged and there cannot be an electric field either because a constant magnetic field is unable to produce an electric field.

The above argument is wrong; don't let yourself be misled. Induction is *not* a local phenomenon. Although $\partial\mathbf{B}/\partial t = 0$ at $x = 0$, it does *not* follow that the electric field is also zero at $x = 0$; the curl of the electric field must be zero but *not* the electric field itself. When the magnetic field is moved bodily there will be certain places in space where the magnitude of the magnetic field is changing (shaded areas in Fig. 4.12) as a function of time, and that changing magnetic field can give rise to an electric field at $x = 0$. The electric field will then produce just the right force for displacing the electrons by the right amount.

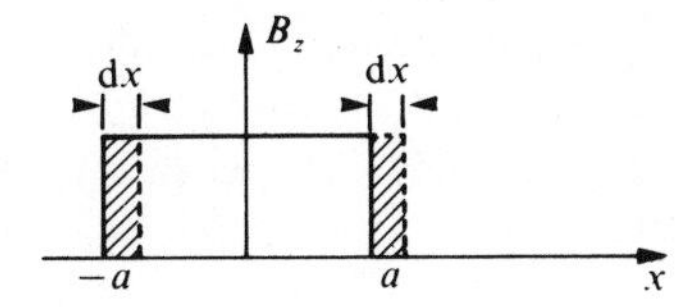

FIG. 4.12. Schematic representation of a moving magnetic field.

Let's attempt to obtain a general solution of this problem. We shall assume a static magnetic field

$$\mathbf{B} = \mathbf{B}(\mathbf{r}), \tag{4.76}$$

which will vary as

$$\mathbf{B} = \mathbf{B}(\mathbf{r} - \mathbf{u}t) \tag{4.77}$$

when moved bodily by a velocity u. The equation to solve is eqn (4.3) which we shall write here again:

$$\nabla \times \mathbf{E} = -\frac{\partial \mathbf{B}}{\partial t}. \tag{4.3}$$

By differentiating eqn (4.77) we get

$$\frac{\partial \mathbf{B}}{\partial t} = -(\mathbf{u} \cdot \nabla)\mathbf{B}. \tag{4.78}$$

But according to eqn (A.7)

$$\begin{aligned} \nabla \times (\mathbf{u} \times \mathbf{B}) &= \mathbf{u}\, \nabla \cdot \mathbf{B} - \mathbf{B}\, \nabla \cdot \mathbf{u} + (\mathbf{B} \cdot \nabla)\mathbf{u} - (\mathbf{u} \cdot \nabla)\mathbf{B} \\ &= -(\mathbf{u} \cdot \nabla)\mathbf{B}, \end{aligned} \tag{4.79}$$

thus the solution of eqn (4.3) may be recognized as being given by

$$\mathbf{E} = -\mathbf{u} \times \mathbf{B}. \tag{4.80}$$

We are now entitled to write the force on the charges as

$$\mathbf{F} = q\mathbf{E} = -q(\mathbf{u} \times \mathbf{B}). \tag{4.81}$$

If the magnetic field is moved with a velocity $-\mathbf{v}$ the force on the charges is the same as if the wire moved with a velocity $\mathbf{v}$, and we have managed to prove all that from Maxwell's equations.

4.7. Faraday's law

In order to derive the general form of Faraday's law we need to notice that motion of charge along the wire may be caused both by electric and magnetic fields. Hence for the calculation of e.m.f. in a moving loop we should take both effects into account leading to a definition in terms of the force per unit charge as follows:

$$\mathscr{E}=\oint\frac{\mathbf{F}}{q}\,.\,\mathrm{d}\mathbf{s}=\oint(\mathbf{E}+\mathrm{v}\times\mathbf{B})\,.\,\mathrm{d}\mathbf{s}=\int\{\nabla\times(\mathbf{E}+\mathbf{v}\times\mathbf{B})\}\,.\,\mathrm{d}\mathbf{S}$$

$$=\int\left\{-\frac{\partial\mathbf{B}}{\partial t}+\nabla\times(\mathbf{v}\times\mathbf{B})\right\}.\,\mathrm{d}\mathbf{S} \tag{4.82}$$

Making use of eqn (4.79) and of the vector derivative relation

$$\frac{\mathrm{d}\mathbf{B}}{\mathrm{d}t}=\frac{\partial\mathbf{B}}{\partial t}+(\mathbf{v}\,.\,\nabla)\mathbf{B}, \tag{4.83}$$

we get finally

$$\mathscr{E}=-\frac{\mathrm{d}\Phi}{\mathrm{d}t}. \tag{4.84}$$

This is the general form of Faraday's law where the derivative is now taken in the frame moving with the wire.

4.8. Flux cutting

Let us take a loop immersed in a magnetic field as shown in Fig. 4.13(a) and assume that a section of the loop moves a distance **dl** coming to a position shown in Fig. 4.13(b) after a time dt. This elementary piece of wire moving with a velocity $\mathbf{dl}/\mathrm{d}t$ contributes to the e.m.f. the amount

$$\left(\frac{\mathrm{d}\mathbf{l}}{\mathrm{d}t}\times\mathbf{B}\right).\,\mathrm{d}\mathbf{s}, \tag{4.85}$$

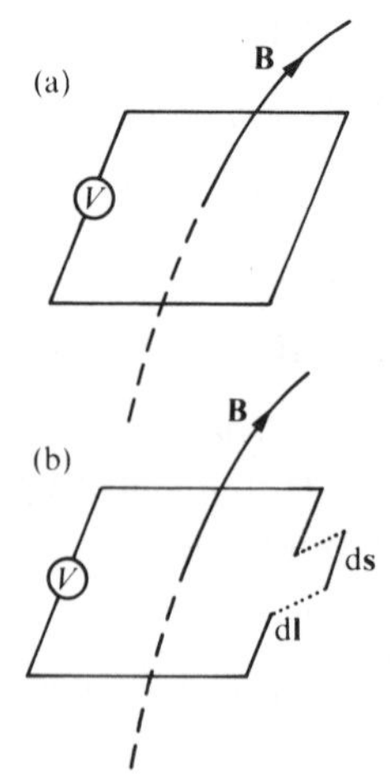

FIG. 4.13. (a) A loop immersed into a magnetic field. (b) Part of the loop moving in the magnetic flux.

which may be rewritten in the form

$$-\frac{d}{dt}\mathbf{B}\cdot(\mathbf{dl}\times\mathbf{ds}) = -\frac{d}{dt}\Delta\Phi, \tag{4.86}$$

where $\Delta\Phi$ is the amount of flux cut by the moving part of the loop. If several distinct sections of the conducting loop are in motion then the flux cut by each section needs to be summed algebraically (each contribution taken as positive or negative depending whether the motion of that section increases or decreases the flux linkage).

The conclusion is that eqn (4.84) is valid under a new set of conditions. The change of flux may be interpreted as the amount of flux cut by the moving sections of the loop.

4.9. Examples on wires moving in a magnetic field

In the following examples we shall use Faraday's law for calculating the e.m.f. in moving loops.

Let us first take a static one-dimensional magnetic vector field pointing in the z direction and varying sinusoidally in the y direction as

$$B_z = B_0 \sin ky, \tag{4.87}$$

where B_0 and k are constants. Assume that a loop lying in the x, y plane (dimensions shown in Fig. 4.14) moves with a uniform velocity v in the

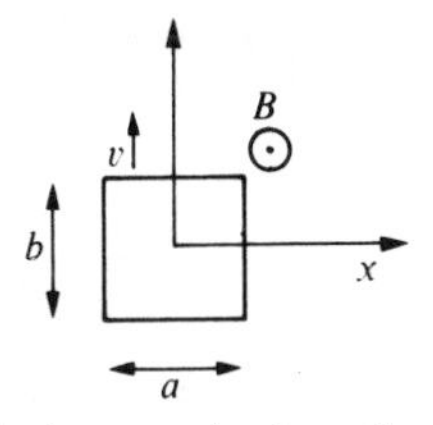

FIG. 4.14. Wire loop moving in static magnetic field.

direction of the positive y axis. For simplicity we shall further assume that the length of the loop in the y direction is smaller than the period of the magnetic field, $b<2\pi/k$. If the rear section of the loop is situated at $y=-b/2$ at time $t=0$, then the positions of the rear and front sections will be

$$y_r = vt - b/2 \quad \text{and} \quad y_f = vt + b/2 \tag{4.88}$$

at time t. The flux enclosed is then

$$\Phi = B_0 a \int_{y_r}^{y_f} \sin ky \, dy = \frac{2B_0 a}{k} \sin\frac{kb}{2} \sin kvt \tag{4.89}$$

and the resulting e.m.f. is

$$\mathscr{E} = -2B_0 av \sin\frac{kb}{2}\cos kvt. \tag{4.90}$$

For our second example we shall take a magnetic field varying sinusoidally both in space and in time:

$$B_z = B_0 \sin ky \sin \omega t. \tag{4.91}$$

Assuming the same loop travelling with the same velocity as in the previous example, the flux through the loop varies as

$$\Phi = \frac{2B_0 a}{k}\sin\frac{kb}{2}\sin kvt \sin \omega t, \tag{4.92}$$

yielding

$$\mathscr{E} = -2B_0 a \sin\frac{kb}{2}\left(kv \cos kvt \sin \omega t + \omega \cos \omega t \sin kvt\right). \tag{4.93}$$

Let us consider now a somewhat different example where instead of uniform translation the loop rotates in a constant magnetic field. This may occur in the same physical configuration as that of Fig. 3.16 (p. 81) used for calculating the torque upon a single turn of current-carrying wire.

Assuming that $\theta = 0$ at $t = 0$ the angular position of the loop is given by $\theta = \omega t$ and the flux across the loop may be obtained as

$$\Phi = B_0 \,.\, ab \cos \omega t, \tag{4.94}$$

whence the e.m.f. is

$$\mathscr{E} = B_0 \,.\, ab\omega \sin \omega t. \tag{4.95}$$

Are you quite happy with this result? I suppose you are. A single turn rotating in a constant magnetic field is one of the standard examples of the application of Faraday's law. The result is used for the design of a.c. generators so it must be all right. Nevertheless, just a tiny little doubt should lurk somewhere at the back of your mind. In this example we are not concerned with uniform translation of the loop but with rotation. The loop is subjected to acceleration; there is a centripetal force acting upon the electrons inside the wire. The present approach is permissible only under the condition that the centripetal force is negligible in comparison with the $e\mathbf{v}\times\mathbf{B}$ force.

The ratio of the two forces may be expressed as

$$\frac{mv\omega}{evB} = \frac{\omega}{\omega_c}, \tag{4.96}$$

where $\omega_c=(e/m)B$ is the so-called *cyclotron frequency*. Thus effects of the acceleration are negligible as long as $\omega \ll \omega_c$. In a practical case (say) $B=1$ T and the loop rotates at 3000 revolution per minute, yielding

$$\frac{\omega}{\omega_c}=\frac{2\pi\times 50}{1{\cdot}76\times 10^{11}}=1.78\times 10^{-9}, \tag{4.97}$$

so we have a safe margin. Nevertheless, keep in mind that results derived for stationary or uniformly moving bodies will not necessarily apply when the body is accelerated.

The concept of flux cutting could have equally been used in the calculations involving static flux.

In our first example where the magnetic field varies according to eqn (4.87) the flux cut by the front part of the moving loop in a time $\mathrm{d}t$ is

$$\mathrm{d}\Phi_f=B_z(y+b/2)av\,\mathrm{d}t, \tag{4.98}$$

and the amount cut by the rear part is

$$\mathrm{d}\Phi_r=B_z(y-b/2)av\,\mathrm{d}t. \tag{4.99}$$

Hence the net flux cut is

$$\mathrm{d}\Phi=\mathrm{d}\Phi_f-\mathrm{d}\Phi_r=avB_0\{\sin k(vt+b/2)-\sin k(vt-b/2)\}\,\mathrm{d}t, \tag{4.100}$$

yielding

$$\mathscr{E}=-2B_0av\sin\frac{kb}{2}\cos kvt, \tag{4.101}$$

in agreement with eqn (4.90).

4.10. Eddy currents

In this section the effect of an e.m.f. on a solid piece of conducting body is studied. In response to the e.m.f. a current (an *eddy current*) flows which will, in general, take very complicated paths. We shall now take a very simple case when a conducting disc of thickness δ, radius b, and of conductivity σ is placed in the magnetic field of eqn (4.19) so that the axes coincide. For simplicity we shall further assume that $b<a$ that is the disc is completely immersed into the magnetic flux.

In the absence of the disc the electric field is given by eqn (4.22). Assuming that the magnetic field produced by the currents in the disc is negligible in comparison with the magnetic field postulated, the electric field distribution remains the same and the corresponding current density is given by

$$J_\varphi=\sigma E_\varphi=-\frac{\sigma B_0\omega}{2}R\cos\omega t. \tag{4.102}$$

At a radius R the current flows in the azimuthal direction in a tube of cross-section $\delta\, \mathrm{d}R$. the average power dissipated by this current is

$\frac{1}{2}J_\varphi^2 \times (\text{cross-section})^2 \times (\text{resistance of the tube})$

$$=\frac{1}{2}\left(\frac{\sigma B_0 \omega R}{2}\right)^2 (\delta\, \mathrm{d}R)^2 \frac{2\pi R}{\sigma\delta\, \mathrm{d}R} = \frac{\pi}{4}\delta\sigma B_0^2 \omega^2 R^3\, \mathrm{d}R. \tag{4.103}$$

The total power dissipated in the disc may then be obtained by integration as follows

$$P = \frac{\pi}{4}\delta\sigma B_0^2 \omega^2 \int_0^b R^3\, \mathrm{d}R = \frac{\pi}{16}\delta\sigma B_0^2 \omega^2 b^4. \tag{4.104}$$

How could we reduce the eddy-current losses in the disc? First, we have to choose a low-conductivity material. Secondly, we could interrupt the currents by insulating various parts of the disc from each other. These problems do indeed arise whenever the magnetic cores of coils and transformers are made of conducting materials (mostly iron). Both remedies suggested above are put into practice. The iron is made into a high-resistivity material by adding silicon, and its cross-section is laminated as shown in Fig. 4.15.

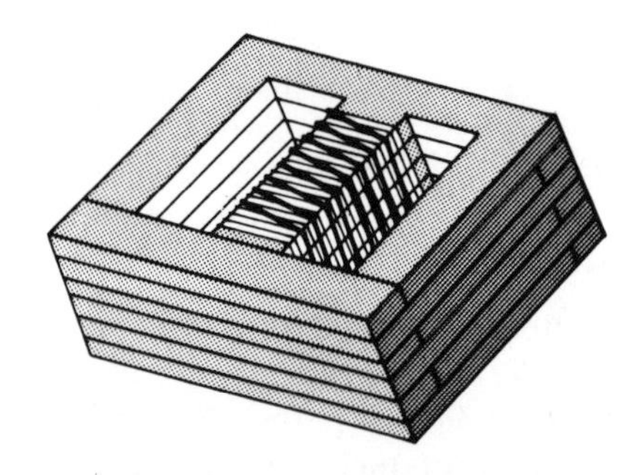

FIG. 4.15. Transformer with laminated core. Both coils wound on the middle section.

4.11. Electromotive force produced by rotating disc

We shall now consider the case when the solid conducting bodies are in motion, the simplest example being when a disc is rotated in a constant magnetic field as shown in Fig. 4.16(a). There will now be a $q(\mathbf{v}\times\mathbf{B})$ radial force on the charges resulting in charge separation and in the appearance of an electric field.

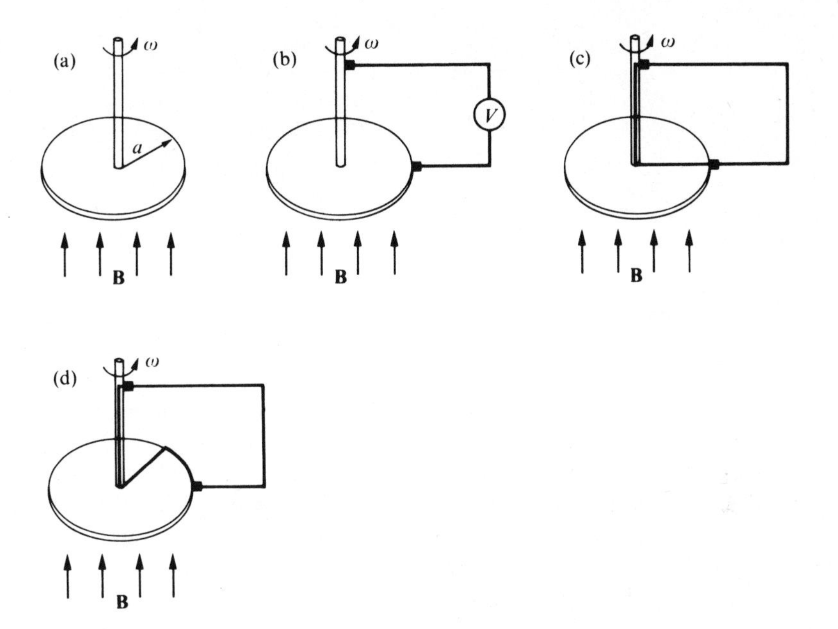

FIG. 4.16. Rotating disc in a magnetic field.

Let us complete now the circuit with the aid of wires and brushes (Fig. 4.16(b)) and insert an ideal voltmeter. What is the voltage measured? It is equal numerically to the e.m.f. in the closed circuit. Since the wires are stationary only the disc contributes to the e.m.f., hence the integration over the closed path reduces to integration along the radius of the disc, yielding

$$\mathscr{E} = \int_0^a \mathbf{v} \times \mathbf{B} \cdot d\mathbf{s} = \omega B_0 \int_0^a r \, dr = \omega B_0 \frac{a^2}{2}. \tag{4.105}$$

Can we work out the e.m.f. with the aid of the concepts of flux linkage or flux cutting? Not easily. If we choose the circuit in the form shown in Fig. 4.16(c) there is neither flux linkage nor flux cutting. However, if we choose our loop as shown in Fig. 4.16(d), where the boundary moves with the rotating disc, then both flux linkage and flux cutting make sense and lead to the desired result.

Let us now make the problem a little more complicated by assuming that only part of the disc is in an appreciable magnetic field, as shown in Fig. 4.17(a). What happens if we rotate the disc? An e.m.f. will duly appear.

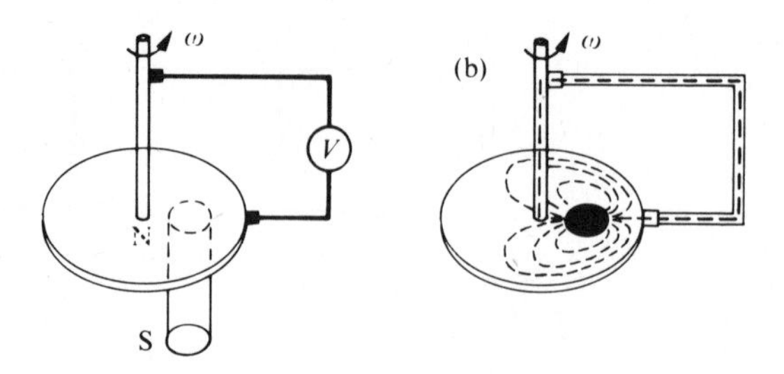

FIG. 4.17. Rotating disc in a partial magnetic field. The area where the magnetic flux cuts the disc is shown by shading. The dotted lines represent current flow.

Why? In the region permeated by the magnetic field the charges are subjected to radial forces and as a consequence eddy currents will flow. The resulting current distribution would be terribly difficult to calculate, but one can make a rough guess and claim that it will flow along the dotted lines shown in Fig. 4.17(b). Omitting the voltmeter from our circuit, so that a current can flow in the resistive wire, it may be seen that some of the current will be forced through the wire hence the device works as a generator.

4.12. Energy and forces

The force acting on a circuit placed in a magnetic field is given by eqn (3.90). If we have (say) two circuits then the forces upon each other may be calculated by determining first the magnetic field over all space and then applying eqn (3.90). An alternative method is to derive first the energy of a circuit (or circuits) and then use the principle of virtual work to find the force. In this section we shall have a brief look at the latter method.

How can we find the energy of a current-carrying circuit? The answer may be easily obtained with the aid of circuit theory. If we start with the initial condition that $I=0$ at $t=0$ and at a later time t the current rises to I then the work done on the inductor is

$$W=\int_0^t VI\,\mathrm{d}t=L\int_0^t I\frac{\mathrm{d}I}{\mathrm{d}t}\mathrm{d}t=L\int_0^I I\,\mathrm{d}I=\tfrac{1}{2}LI^2. \tag{4.106}$$

The same result may be derived from field theory in the following manner. If a current element $I\,\mathrm{d}\mathbf{s}$ is displaced parallel to itself a distance $\mathrm{d}\mathbf{l}$ in a magnetic flux density $\mathbf{B}$ then the work done is

$$\mathrm{d}W=\mathbf{F}\,.\,\mathrm{d}\mathbf{l}=I(\mathrm{d}\mathbf{s}\times\mathbf{B})\,.\,\mathrm{d}\mathbf{l}=I\mathbf{B}\,.\,(\mathrm{d}\mathbf{l}\times\mathrm{d}\mathbf{s})=I\,\mathrm{d}\Phi=LI\,\mathrm{d}I, \tag{4.107}$$

leading again to $W=\tfrac{1}{2}LI^2$.

For two coupled inductors a similar derivation (either by circuit or by field theory) yields

$$W=\tfrac{1}{2}L_1I_1^2+MI_1I_2+\tfrac{1}{2}L_2I_2^2. \tag{4.108}$$

Let us rewrite the above equation in the following form

$$W = \frac{1}{2} L_1 \left(I_1 + \frac{M}{L_1} I_2 \right)^2 + \frac{1}{2} \left(L_2 - \frac{M^2}{L_1} \right) I_2^2. \tag{4.109}$$

Since the stored energy must always be positive for any pair of I_1 and I_2, we may choose

$$I_2 = -\frac{L_1}{M} I_1, \tag{4.110}$$

whence it follows that the condition

$$M^2 \leqslant L_1 L_2 \tag{4.111}$$

is valid for any two coupled inductors.

Let us take now two rigid loops in which constant currents I_1 and I_2 flow. The total stored energy is given by eqn (4.108). What is the change in energy if we move one of the loops by a distance dx? The self-inductances will stay constant (because the loops are rigid), but the mutual inductance depends on the distance between the current elements so it will change by an amount dM. Thus the change in stored energy comes to

$$dW = I_1 I_2 \, dM, \tag{4.112}$$

whence one could conclude that

$$F = -\frac{dW}{dx} = -I_1 I_2 \frac{dM}{dx}.$$

This is, of course, wrong as we could have guessed on the basis of our experience with forces acting upon capacitor plates. We need to include the work done by the sources in keeping the currents constant. This amounts to

$$dW_0 = I_1 \, d\Phi_{21} + I_2 \, d\Phi_{12} = 2 I_1 I_2 \, dM. \tag{4.113}$$

Hence

$$F \, dx = -I_1 I_2 \, dM + 2 I_1 I_2 \, dM = I_1 I_2 \, dM. \tag{4.114}$$

As an example let us now work out the forces between two current-carrying rings with the aid of our new formula. The mutual inductance is given by eqn (4.57). The force may then be calculated as

$$F = I_1 I_2 \frac{dM}{dh} = -I_1 I_2 \frac{\mu_0 \pi}{2} a_1^2 a_2^2 \frac{3}{2} \frac{2h}{(a_1^2 + h^2)^{\frac{5}{2}}}, \tag{4.115}$$

in agreement with eqn (3.93).

Examples 4

4.1. A circular coil of diameter 20 mm has 100 turns and lies at the centre of a solenoid for which $s = 0{\cdot}3$ m, $d = 0{\cdot}1$ m, and $n = 800$, the coil and the solenoid being coaxial (s = total length, d = diameter, n = turns per metre). What is the e.m.f. induced in the open-circuited coil if the solenoid is fed with a current of 3A at 50 Hz?

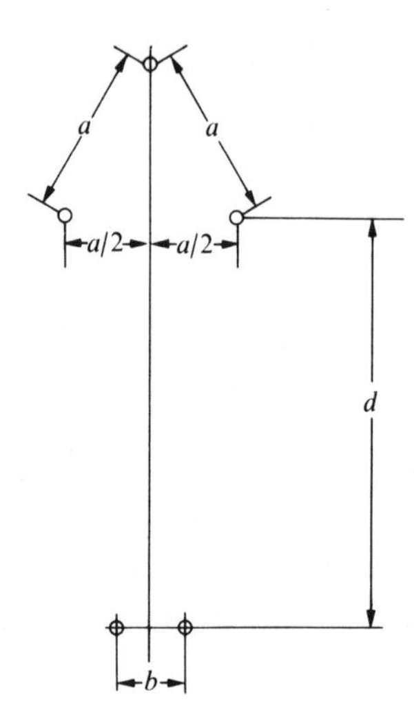

FIG. 4.18. A loop in the vicinity of a three-wire transmission line.

4.2. Fig. 4.18 shows the cross-section of a straight transmission line carrying currents

$$I_1 = I_0 \exp \mathrm{j}\omega t, \qquad I_2 = I_0 \exp \mathrm{j}(\omega t - \tfrac{2}{3}\pi), \qquad I_3 = I_0 \exp \mathrm{j}(\omega t + \tfrac{2}{3}\pi),$$

where ω is the angular frequency and I_0 is a constant.

Derive a formula for the e.m.f. induced around a loop comprising a length l of short-circuited cable running parallel with the transmission line. Assume that $d > a$ and $b \ll a$.

Calculate the e.m.f. when $I_0 = 1000$ A, $a = 1$ m, $b = 3$ mm, $d = 5$ m, $\omega = 100\pi\ \mathrm{s}^{-1}$, $l = 1$ m.

4.3. A long, thin-walled, non-magnetic, conducting tube of radius R, wall-thickness δ, and conductivity σ is placed with its axis parallel to the direction of a uniform alternating magnetic field $H = H_0 \exp \mathrm{j}\omega t$.

Determine the current per unit length flowing in the wall, the magnetic field inside the tube and the average power dissipated per unit length. Neglect edge effects that is assume no change in the variables along the tube.

4.4. Show that the mutual inductance between two arbitrary coils may be written in the form (see Fig. 4.8, p. 100)

$$M_{12} = M_{21} = M = \frac{\mu_0}{4\pi} \oint_1 \oint_2 \frac{\mathrm{d}\mathbf{s}_1\, \mathrm{d}\mathbf{s}_2}{r_{12}}.$$

4.5. Eqn (4.57) gives the mutual inductance of two concentric rings a distance h apart from each other. If a_1 and a_2 are interchanged the mutual inductance will not remain invariant, defying the $M_{12} = M_{21}$ relationship. What is the cause of this discrepancy?

4.6. Determine the mutual impedance between the coil and the solenoid of Example 4.1.

4.7. Eqn (4.57) gives the mutual inductance of two concentric rings a distance h apart from each other. Assume that ring 1 carries a current $I_1 = I_0 \exp j\omega t$ and the resistance and self-inductance of ring 2 are given as L_2 and R_2.

Show that the mean value of the force F between the two rings (i.e. averaged over the periodic time of the alternating current) is given by

$$F = \tfrac{3}{8}|I_1|^2 \frac{\omega^2 L_2}{R_2^2 + (\omega L_2)^2} \frac{h}{(a_1^2 + h^2)^4} (\mu_0 \pi a_1^2 a_2^2)^2,$$

and say which way it acts.

It is desired to maximize the ratio (F/weight of ring), and for this purpose it is necessary to decide whether for any given geometry it is better to make the ring of copper or of aluminium. Is it possible to give an answer without having numerical values for anything except the properties of aluminium and copper?

	Conductivity at 20°C	Density (kg/m^{-3})
Aluminium	$3{\cdot}74 \times 10^7$	2700
Copper	$5{\cdot}8 \times 10^7$	8960

4.8. Assume an axially symmetric radial magnetic flux density in empty space between the radii R_1 and R_2 which is constant in the vertical direction (z). Place now a thin wire of radius $\rho (R_1 < \rho < R_2)$ into the magnetic field at a height z_0 so that its plane is perpendicular to the z-*axis*, and each element of the wire is subjected to a constant radial flux density B_r. At $t = 0$ the ring is released with zero initial velocity,

(i) Derive the equation of motion for the ring and show that it is independent of the cross-section of the wire.

(ii) Determine the position and velocity of the ring as a function of time.

(iii) What is the limiting velocity as $t \to \infty$?

(iv) In which case will the ring fall faster, if it is made of copper or of aluminium?

4.9. A transformer supplied from a 220 V, 50 Hz mains has an iron core in which the peak flux density, 0·66 T, is reached at a magnetic field of 150 A m^{-1}. The cross-section of the core is $1{\cdot}5 \times 10^{-3}$ m^2 and its volume is 9×10^{-4} m^3.

Determine the magnetization current and the number of turns in the primary circuit.

4.10. A rectangular loop moving with a velocity v is shown schematically in Fig. 4.14 (p. 109) and the resulting e.m.f. is given by eqn (4.90). Assume now that the loop is stationary and the magnetic field (given by eqn (4.87)) moves with the same velocity in the opposite direction. Determine the e.m.f. by taking the line integral of the electric field around the loop.

4.11. A rectangular coil is located parallel to a long straight current-carrying wire as shown in Fig. 4.19. Determine the e.m.f. in the coil when it is rotated with an angular velocity ω.

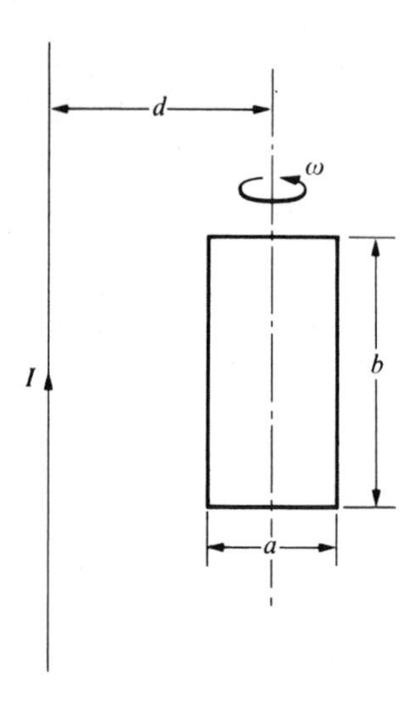

FIG. 4.19. A loop in the vicinity of a straight current carrying wire.

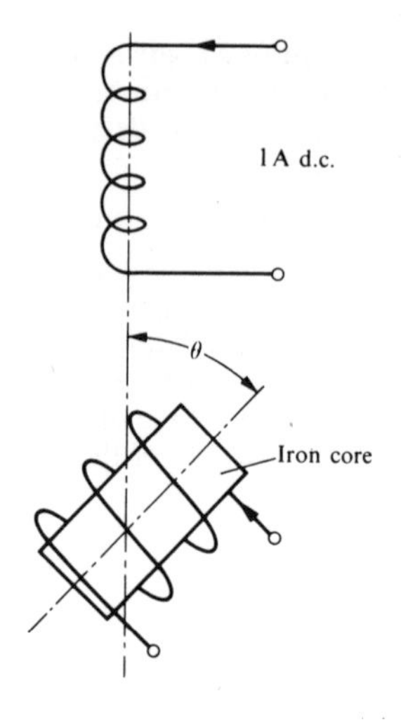

FIG. 4.20. Two coupled coils.

4.12. Take the same geometrical configuration as in the previous example but assume that the rectangular coil is moving with a constant velocity v in a direction perpendicular to the straight wire.

Determine the e.m.f. in the coil.

4.13. Find the torque which tends to align the rotor in the arrangement shown in Fig. 4.20 with $\theta = 45°$ and $135°$, (i) when the rotor carries no current, and (ii) when it carries a direct current of 2 A. The rotor inductance is 1 H and the stator inductance has maximum and minimum values 1 H and 0·2 H. Assume that periodic inductances vary in a sinusoidal manner and that the coils are perfectly coupled when they are in line.

5. Fast-varying phenomena

In a dielectric under the action of electromotive force, we may conceive that the electricity in each molecule is so displaced that one side is rendered positively and the other negatively electrical, but that the electricity remains entirely connected with the molecule, and does not pass from one molecule to another. The effect of this action on the whole dielectric mass is to produce a general displacement of electricity in a certain direction. This displacement does not amount to a current, because when it has attained to a certain value it remains constant, but it is the commencement of a current, and its variations constitute currents in the positive or the negative direction according as the displacement is increasing or decreasing.

The variations of the electrical displacement must be added to the currents to get the total motion of electricity.

If the medium, instead of being a perfect insulator, is a conductor whose resistance per unit of volume is ρ, then there will be not only electric displacements, but true currents of conduction in which electrical energy is transformed into heat, and the undulation is thereby weakened.

The agreement of the results seems to shew that light and magnetism are affections of the same substance, and that light is an electromagnetic disturbance propagated through the field according to electromagnetic laws.

JAMES CLARK MAXWELL *A dynamical theory of the electric field, Phil. Trans.* **155** 1865

When, about two years ago, news came from the other side of the Atlantic that a method had been invented of transmitting, by means of electricity, the articulate sounds of the human voice, so as to be heard hundreds of miles away from the speaker, those of us who had reason to believe that the report had some foundation in fact, began to exercise our imaginations in picturing some triumph of constructive skill. When at last this little instrument appeared, consisting, as it does, of parts, everyone of which is familiar to us, and capable of being put together by an amateur, the disappointment arising from its humble appearance was only partially relieved on finding that it was really able to talk.

I have not yet met anyone acquainted with the first elements of electricity who has experienced the slightest difficulty in understanding the physical process involved in the action of the telephone.

It would be as useless as it would be tedious to try to explain the various parts of this small instrument to persons in every part of the Senate House. I shall, therefore, consider the telephone as a material symbol of the widely separated departments of human knowledge, the cultivation of which has led, by as many converging paths, to the invention of this instrument by Professor Graham Bell.

For whatever may be said about the importance of aiming at depth rather than width in our studies, and however strong the demand of the present age may be for specialists, there will always be work, not only for those who build up particular sciences and write monographs on them, but for those who open up such communications between the different groups of builders as will facilitate a healthy interaction between them. And in a university we are especially bound to recognise not only the unity of science itself, but the communion of the workers in science. We are too apt to suppose that we are congregated here merely to be within reach of certain appliances of study, such as museums and laboratories, libraries and lecturers, so that each of us may study what he prefers.

James Clerk Maxwell *The telephone (Rede Lecture), Nature* **18** 1878

. . . in a medium in which waves are propagated there is a pressure in the direction normal to the waves . . . A flat body exposed to sunlight would experience this pressure on its illuminated side only, and would therefore be repelled from the side on which the light falls. It is probable that a much greater energy of radiation might be obtained by means of the concentrated rays of the electric lamp. Such rays falling on a thin metallic disk, delicately suspended in a vacuum, might perhaps produce an observable mechanical effect.

James Clerk Maxwell *A treatise on electricity and magnetism* Oxford 1873

. . . the existence of induced currents and of electromagnetic actions at a distance from a primary circuit from which they draw their energy, has led us, under the guidance of Faraday and Maxwell, to look upon the medium surrounding the conductor as playing a very important part in the development of the phenomena. If we believe in the continuity of the motion of energy, that is, if we believe that when it disappears at one point and reappears at another it must have passed through the intervening space, we are forced to conclude that the surrounding medium contains at least a part of the energy, and that it is capable of transferring it from point to point.

I think it is necessary that we should realise thoroughly that if we accept Maxwell's theory of energy residing in the medium, we must no longer consider a current as something conveying energy along the conductor. A current in a conductor is rather to be regarded as consisting essentially of a convergence of electric and magnetic energy from the medium upon the conductor and its transformation there into other forms.

J. H. Poynting *On the transfer of energy in the electromagnetic field, Phil. Trans.* **175** 1884

Der Plan, welcher für die Untersuchung aufgestellt wurde, war der folgende: Zuerst sollten mit Hülfe der schnellen Schwingungen eines primären Leiters entsprechende regelmässige, fortschreitende Wellen in einem geradlinig ausgespannten Drahte erzeugt werden. Zuzweit sollte ein secundärer Leiter gleichzeitig der Einwirkung der durch den Draht fortgepflanzten Wellen und der durch die Luft fortgepflanzten directen Wirkung der primären Schwingung ausgesetzt und so beide Wirkungen zur Interferenz gebracht werden. Endlich sollten solche Interferenzen in verschiedenen Abständen vom primären Kreise hergestellt und so ermittelt werden, ob die Schwingungen der elektrischen Kraft in grösseren Entfernungen eine Phasenverzögerung gegen die Schwingungen in der Nähe aufwiesen oder nicht. Ein vorhandener Phasenunterschied würde eine endliche Ausbreitungsgeschwindigkeit anzeigen. Dieser Plan hat sich in allen Theilen als durchführbar erwiesen. Die nach ihm angestellten Versuche haben ergeben, dass sich die Inductionswirkung durch den Luftraum allerdings mit endlicher Geschwindigkeit ausbreitet.

HEINRICH HERZ *Ueber die Ausbreitungsgeschwindigkeit der elektrodynamischen Wirkungen, Wiedemanns Ann.* **34** 1888

Es ist gewiss ein interessanter Gedanke, dass die Vorgänge im Luftraume, welche wir untersuchten, uns in millionenfacher Vergrösserung dieselben Vorgänge darstellen, welche zwischen den Platten eines Newton'schen Farbenglases oder in der Nähe eines Fresnel'schen Spiegels sich abspielen.

HEINRICH HERTZ *Ueber elektrodynamische Wellen im Luftraume und deren Reflexion, Wiedemanns Ann.* **34** 1888

Ich habe die Versuche in der von Maxwell vorgeschlagenen einfachen Form aufgenommen.

Diese Druckkräfte des Lichtes stimmen innerhalb der Versuchsfehler quantitativ mit den von Maxwell und von Bartoli berechneten ponderomotorischen Kräften der Strahlung überein.
Hierdurch ist die Existenz der Maxwell–Bartoli'schen Druckkräfte für Lichtstrahlen experimentell erwiesen.
Moskau, Physik. Laborat. d. Univ., im August 1901.

P. LEBEDEW *Untersuchungen über die Druckkräfte des Lichtes, Ann. Phys.* **4, 6** 1901

PROPOSITIO I.

Lumen propagatur seu diffunditur non solum Directe, Refractè, ac Reflexè, sed etiam alio quodam Quarto modo, Diffracte

Aperto in fenestra foraminulo perquam paruo AB, inroducatur per illud in cubiculum, alioqui valdè obscurum, lumen Solis Cælo serenissimo, cuius diffusio erit per conum, vel quasi conú ACDB visibilem si aër fuerit refertus atomis pulueris, vel si in co excitetur aliquis fumus. Huic cono inferatur aliquod corpus opacum EF, in magna distantia à foramine AB, et ita vt faltem vnum extremum corporis opaci illuminetur.

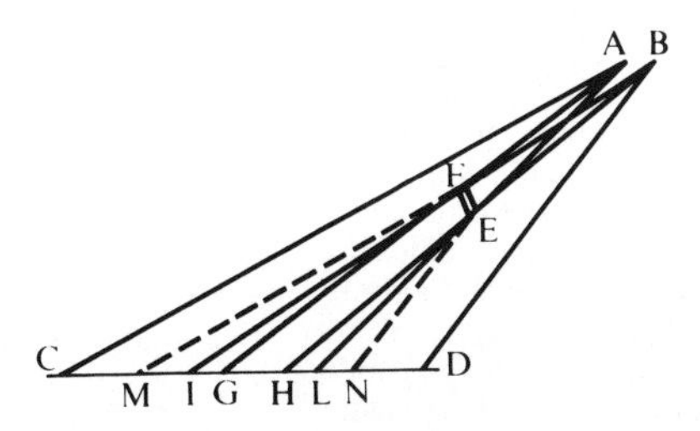

Prætereà obseruetur super lucidæ basis parte CM, et ND, nitidè ac fortiter illustrata, spargi et distingui tractus aliquos, seu series luminis colorati, ita vt in qualibet serie sit in medio quidem lux valdè pura, et sincera, in extremis autem sit color aliquis, nempe cæruleus in extremo ipsi vmbræ MN propinquiore, et rubeus in extremo remotiore:

PROPOSITIO XXII.
Lumen aliquando per sui communicationem reddit obscuriorem superficiem corporis aliunde, ac priùs illustratam.

Hæc Propositio paradoxum est, et ex terminis ipsis magnam præfert improbabilitatem, quia luminis est illustrare, non autem obscura superficiem corporis opaci, ad quá terminatur, et cui aliquo tandem modo se communicat. Eius tamen probatio certissima est, ac euidenter manifesta ex aliquo experimento valde obuio, sed hactenus à nemine, quòd sciam, considerato. Hoc igitur breuiter priùs exponendum est.

FRANCISCO MARIA GRIMALDO *Physico-mathesis de Lumine, coloribus, et iride, alliisque adnexis* Bononiae MDCLXV

This Learned Treatise was not to be altogether omitted in those Philosophical Occurences, though an Account of it hath been deferr'd (too long), it being but lately fallen into the Publisher's hands.
The Author then finding, that much obscurity was left in the Doctrine of Light, and esteeming it rather commendable than presumptuous, to endeavor the clearing of it, especially if that be done by Experiments (which he judgeth an exceedingly conducive way for the Improvement of All Natural Knowledge;) undertaketh in Two parts to deliver his Tryals and Meditations on this Subject.

But more particularly, in the former part he explains, How many ways Light is propagated or diffused, viz. not only directly, and by refraction, and reflexion, but also by diffraction; which last according to him, is done when the parts of Light, separated by a manifold dissection, do in the same medium proceed in different ways.

Review of Grimaldi's *De lumine* in *Phil. Trans.* **6** 1672

I'escrivis ce Traité pendant mon sejour en France, il y a 12 ans; et je le communiquay en l'année 1678 aux personnes sçavantes, qui composoient alors l'Academie Royale des Sciences, à la quelle le Roy m'avoit fait l'honneur de m'appeller

On pourra demander pourquoy j'ay tant tardé à mettre au jour cet Ouvrage. La raison est que je l'avois escrit assez negligemment en la Langue où on le voit avec intention de le traduire en Latin faisant ainsi pour avoir plus d'attention aux choses. Apres quoy je me proposois de le donner ensemble avec un autre Traité de Dioptrique, ou j'explique les effets des Telescopes, et ce qui apartient de plus à cette Science. Mais le plaisir de la nouveauté ayant cessé, j'ay differé de temps à autre d'executer ce dessein, et je ne scay pas quand j'aurois encore pû en venir à bout, estant souvent diverti, ou par des affaires, ou par quelque nouvelle étude. Ce que considerant, j'ay en fin jugé qu'il valoit mieux de faire paroitre cet escrit tel qu'il est, que de le laisser courir risque, en attendant plus longtemps, de demeurer perdu.

Ainsi ce nombre infini d'ondes qui naissent en mesme instant de tous les points d'une étoile fixe, grande peut estre comme le Soleil, ne font sensiblement qu'une seule onde, laquelle peut bien avoir assez de force pour faire impression sur nos yeux. Outre que de chaque point lumineux il peut venir plusieurs miliers d'ondes dans le moindre temps imaginable, par la frequente percussion des corpuscules, qui frappent l'Ether en ces points; ce qui contribue encore à rendre leur action plus sensible.

Il y a encore à considerer dans l'émanation de ces ondes, que chaque particule de la matiere, dans laquelle une onde s'etend, no doit pas communiquer son mouvement seulement à la particule prochaine, qui est dans la ligne droite tireé du point lumineux; mais qu'elle en donne aussi necessairement à toutes les autres qui la touchent, et qui s'opposent à son mouvement. De sorte qu'il faut qu'autour de chaque particule il se fasse une onde dont cette particule soit le centre.

CHRISTIAAN HUYGENS *Traité de la lumière* Leyden 1690

'Are not all hypotheses erroneous in which light is supposed to consist in pression or motion propagated through a fluid medium? If it consisted in pression or motion, propagated either in an instant, or in time, it would bend into the shadow. For pression or motion cannot be propagated in a fluid in right lines beyond an obstacle which stops part of the motion, but will bend and spread every way into the quiescent medium which lies beyond the obstacle. The waves on the surface of stagnating water passing by the sides of a broad obstacle which stops part of them, bend afterwards, and dilate themselves gradually into the quiet water behind the obstacle. The waves, pulses, or vibrations of the air, wherein sounds consist, bend manifestly, though not so much as the waves of water. For a bell or a cannon may be heard beyond a hill which intercepts the sight of the sounding body; and sounds are propagated as readily through crooked pipes as straight ones. But light is never known to follow crooked passages nor to bend into the shadow. For the fixed stars, by the interposition of any of the planets, cease to be seen. And so do the parts of the sun by the interposition of the moon, Mercury, or Venus. The rays which pass very near to the edges of any body are bent a little by the action of the body; but this bending is not towards but from the shadow, and is performed only in the passage of the ray by the body, and at a very small distance from it. So soon as the ray is past the body it goes right on.'

ISAAC NEWTON *Opticks* 2nd edition, 1717

Although the invention of plausible hypotheses, independent of any connection with experimental observations, can be of very little use in the promotion of natural knowledge; yet the discovery of simple and uniform principles, by which a great number of apparently heterogeneous phenomena are reduced to coherent and universal laws, must ever be allowed to be of considerable importance towards the improvement of the human intellect.

The object of the present dissertation is not so much to propose any opinions which are absolutely new, as to refer some theories, which have been already advanced, to their original inventors, to support them by additional evidence, and to apply them to a great number of diversified facts, which have hitherto been buried in obscurity.

. . . yet some facts, hitherto unobserved, will be brought forwards, in order to show the perfect agreement of that system with the multifarious phenomena of nature.

Let the concentric lines in Fig. 1† represent the contemporaneous situation of similar parts of a number of successive undulations diverging from the point A; they will also represent the successive situations of each individual undulation: let the force of each undulation be represented by the breadth of the line, and let the cone of light ABC be admitted through the aperture BC; then the principal undulations will proceed in a rectilinear direction towards GH, and the faint radiations on each side will diverge from B and C as centres,

The Bakerian Lecture. On the Theory of Light and Colours. By Thomas Young, *M.D.F.R.S. Professor of Natural Philosophy in the Royal Institution.*

Read November 12, 1801.

Published in *Phil. Trans.* **20** 1802

The experiment of Grimaldi on the crested fringes within the shadow, together with several others of his observations equally important, has been left unnoticed by Newton. Those who are attached to the Newtonian theory of light, or to the hypothesis of modern opticians founded on views still less enlarged, would do well to endeavor to imagine anything like an explanation of these experiments derived from their own doctrines; and if they fail in the attempt, to refrain at least from idle declamation against a system which is founded on the accuracy of its application to all these facts, and to a thousand others of a similar nature.

The observations on the effects of diffraction and interference may, perhaps, sometimes be applied to a practical purpose in making us cautious in our conclusions respecting the appearances of minute bodies viewed in a microscope.

THOMAS YOUNG *Bakerian Lecture* 1803

I have also been reflecting on the possibility of giving an imperfect explanation of the affection of light which constitutes polarisation, without departing from the genuine doctrine of undulations. It is a principle, in this theory, that all undulations are simply propagated through homogeneous mediums in concentric spherical surfaces, like the undulations of sound, consisting simply in the direct and retrograde motions of the particles in the direction of the radius, with their concomitant condensation and rarefactions. And yet it is possible to explain in this theory a transverse vibration, propagated also in the direction of the radius, and with equal velocity, the motions of the particles being in a certain constant direction with respect to that radius; and this is a *polarisation.*

From Young's letter to Arago, 12th January, 1817

La première hypothèse qui se présente á la pensée, c'est qu'elles sont produites par la rencontre des rayons directs et des rayons réfléchis sur les bords du corps opaque . . . Telle paraît être l'opinion de M. Young, et c'est aussi celle que j'avais adoptée d'abord, avant qu'un examen plus approfondi des phénomènes m'en eut fait reconnaître l'inexactitude.

Après avoir démontré dans la première section de ce Mémoire, que le système de l'émission, et même le principe des interférences, quand on ne l'applique qu'aux rayons directs et aux rayons réfléchis ou

† See Fig. 5.39a.

infléchis sur les bords mêmes de l'écran, sont insuffisants pour expliquer les phénomènes de la diffraction, je vais faire voir maintenant qu'on peut en donner une explication satisfaisante et une théorie générale, dans le système des ondulations, sans le secours d'aucune hypothese secondaire, et en s'appuyant seulement sur le principe d'Huyghens et sur celui des interférences, qui sont l'un et l'autre des consequences de l'hypothèse fondamentale.

AUGUSTIN FRESNEL *Memoire sur la diffraction* 1819

You will imagine how greatly I have been interested with the two principal papers in the *Annales de chimie* for May. Perhaps, indeed, you will suspect that I am not a little provoked to think that so immediate a consequence of the Huyghenian system, as that which Mr. Fresnel has very ingeniously deduced, should have escaped myself, when I was endeavouring to apply it to the phenomena in question: but in fact, I am still at a loss to understand the possibility of the thing.

From Young's letter to Arago, 4th August, 1819

J'ai l'honneur de vous adresser deux exemplaires de mon Mémoire sur la diffraction, tel qu'il vient d'être imprimé dans les *Annales de chimie et de physique.*

L'extrait publié contient la partie essentielle de mon Mémoire: la théorie de la diffraction et sa vérification expérimentale. Cette théorie, comme vous l'avez très-bien dit, n'est autre chose que le principe d'Huyghens appliqué aux phénomènes en question.

Le principe d'Huyghens me paraît, aussi bien que celui des interférences, une conséquence rigoureuse de la coexistence des petits mouvements dans les vibrations des fluides. Une onde dérivée peut être considérée comme l'assemblage d'une infinité d'ébranlements simultanés; on peut donc dire, d'après le principe de la coexistence des petits mouvements, que les vibrations excitées par cette onde dans un point quelconque du fluide situé au delà sont la somme de toutes les agitations qu'y aurait fait naître chacun de ces centres d'ébranlement en agissant isolément.

From Fresnel's letter to Young, 19th September, 1819

Je vous remercie infiniment, Monsieur, pour le présent que vous m'avez fait de votre beau Mémoire, qui mérite assurément un rang distingué parmi les écrits qui ont le plus contribué aux progrès de l'optique. Je n'ai pas la moindre idée d'insister sur l'opération des rayons réfléchis des bords d'un corps opaque;

From Young's letter to Fresnel, 16th October, 1819

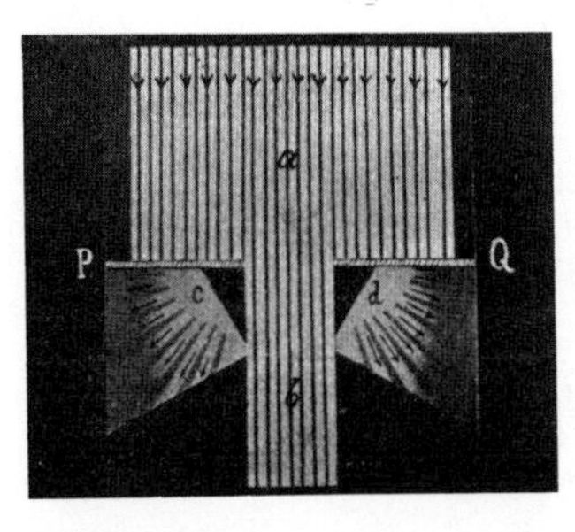

Болѣе внимательное наблюденіе, какъ мы видѣли, обнаруживаетъ, однако, у самаго края экрана очень слабое разсѣяніе свѣта во всѣ стороны. Это видно на снимкѣ 13, гдѣ представленъ ходъ лучей, скользящихъ по поверхности фотографической пластинки, на которой была поставлена преграда *PQ*. Такъ какъ разсѣянный свѣтъ, какъ показываетъ подсчетъ по времени экспозиціи, въ милліоны разъ слабѣе прямого свѣта, то части c и d пластинки, гдѣ замѣтно его присутствіе, должны быть экспонированы въ милліоны разъ дольше средней части a и b. Поэтому такой снимокъ можетъ быть полученъ только искусственнымъ сложеніемъ трехъ снимковъ, изъ которыхъ средній приготовленъ отдѣльно отъ двухъ боковыхъ, полученныхъ по описанному выше способу. Съ точки зрѣнія принципа Гюйгенса это явленіе можно разсматривать, какъ лучеиспусканіе вторичныхъ свѣтящихся точекъ, расположенныхъ у самаго края экрана.

A. Kalashnikow *Gouy–Sommerfeld diffraction, Journal of the Russian Physical and Chemical Society* **44** 1912

Gleichwohl ist es interessant, zu sehen daβ die Youngsche Theorie vom leuchtenden Schirmrande sich nicht nur zufällig bestätigt. Wie wir nämlich zeigen wollen, läβt sich die Funktion durch eine einfache Umformung spalten in eine im Sinne der geometrischen Optik zu definierende *einfallende Lichtwelle* und in eine vom Schirmrande ausgehende *Beugungswelle.* Dabei kann man sich die Beugungswelle nach einem Elementargesetze entstanden denken, welches jedem Punkte des Schirmrandes eine, allein durch die geometrische Lage des betreffenden Randelementes und der Lichtquelle bestimmte, unsymmetrische Kugelwelle zuordnet.

A. Rubinowitz *Annalen der Physik* **53** 1917

Wenn wir hier mit Thomas Young von einer Reflexion des einfallenden Lichtes am Schirmrande sprechen wollen, so ist die 'Art der Reflexion' eine sehr spezialisierte und muβ genau definiert werden.

Es entsteht die Frage, ob und wie sich die Youngsche Auffassung auf beliebige Beugungsschirme erweitern läβt.

Arnold Sommerfeld *Vorlesungen über theoretische Physik, Band IV Optik* Wiesbaden 1950

The cause of diffraction effects was thus wholly unknown, until Young, in the Bakerian lecture for 1803, showed that the principle of interference is concerned in their formation;

His conjecture as to the origin of the interfering rays was not so fortunate; for he attributed the fringes outside the geometrical shadow to interference between the direct rays and rays reflected at the diffracting edge; and supposed the internal fringes of the shadow of a narrow object to be due to the interference of rays inflected by the two edges of the object.

On 15 July 1816 for the first time, diffraction-effects are referred to their true cause—namely, the mutual interference of the secondary waves emitted by those portions of the original wave-front which have not been obstructed by the diffracting screen. Fresnel's method of calculation utilised the principles of both Huygens and Young; he summed the effects due to different portions of the same primary wave-front.

EDMUND WHITTAKER *A history of the theories of aether and electricity* second edition London 1951

Diffraction of a wave by an aperture of any shape in a thin screen is treated by a new method—'the geometrical theory of diffraction.' This is an extension of geometrical optics which accounts for diffraction by introducing new rays called diffracted rays. They are produced when incident rays hit the aperture edge and they satisfy the 'law of diffraction.' A field is associated with each ray in a quantitative way, by means of the optical principles of phase variation and energy conservation. In addition 'diffraction coefficients' are introduced to relate the field on a diffracted ray to that on the corresponding incident ray.

J. B. KELLER *Journal of Applied Physics* **28** 1957

5.1. Introduction

WE have reached the end of the road, at least as far as the number of terms in Maxwell's equations are concerned. In the present chapter we have all of them, so one might draw the tentative conclusion that the present chapter is going to be more difficult than the foregoing ones. Well, there will be a few topics which might require somewhat different mathematics, but I wouldn't call them more difficult. Fast-varying phenomena are by no means more complicated than other parts of electromagnetic theory. I believe it is mainly lack of familiarity, nurtured carefully by your teachers in secondary schools, that breeds fear. An electromagnetic wave is conceptually simple besides being firmly rooted in everyday experience. Whenever you switch on your TV set you rely on the propagation of electromagnetic waves. Reflection and refraction are easy concepts too. The trouble might be that when you were taught optics it may not have been sufficiently emphasized that light waves obey the same laws as electromagnetic waves, that light *is* electricity. Guidance of electromagnetic waves by metal pipes or dielectrics is again a topic that needs no more than common sense provided you are willing to think about it.

I know many teachers of electricity who abhor fast-varying (high-frequency) phenomena, regard them as being in a separate class, unsuitable for mass consumption. They think of waveguides as solutions of Maxwell's equations subject to certain boundary conditions, instead of regarding them as the natural carriers of electricity. Anyone who is not blind, and I mean it literally, anyone who can see the rays of the sun, should find the propagation of electricity in hollow tubes self-evident. Anyone who met the concept of resonance in mechanics should welcome microwave cavities and optical open resonators as natural embodiments of electric resonators and should look upon *LC* circuits as curious contraptions devised by narrow specialists in some very 'electric' branch of electricity.

To my mind high frequencies are nearer to common sense than low frequencies, but that is not the main reason for their increasing popularity. The reason is more to be sought in our insatiable appetite for communication (more and more frequency bands are needed) and in our passionate yearning for speed. We want faster motor cars, faster aeroplanes, faster computers. Fast computers need to work with short pulses and a short pulse (say 0·1 ns duration) has to be treated with all the respect due to fast-varying phenomena. So, besides radar and communications, computers are also turning towards higher frequencies.

It is difficult for low-frequency people to appreciate the pure delights and dizzy heights of high frequencies. High frequencies appear to them as cold unapproachable mountain peaks inhabited by people of a different mould worshipping different gods. The highlanders, on the other hand, know everything about the people in the valleys, even the origin of their superstitions.

'Look', a highlander would say to another highlander, 'they use two wires in order to concentrate the fields into a small volume, and then they believe', he would add with an ironic smile, 'that electricity must go to and fro. They admit the rays of sun but they don't identify them with electricity because their consumption of sunlight does not figure in their electricity bills.'

5.2. The basic equations

We shall now need all eqns (1.1) to (1.7) (p. 5). For easier reference we shall reproduce them below†

$$\nabla \times \mathbf{H} = \mathbf{J} + \frac{\partial \mathbf{D}}{\partial t}, \tag{5.1}$$

$$\nabla \times \mathbf{E} = -\frac{\partial \mathbf{B}}{\partial t}, \tag{5.2}$$

$$\nabla \cdot \mathbf{D} = \rho, \tag{5.3}$$

$$\nabla \cdot \mathbf{B} = 0, \tag{5.4}$$

$$\mathbf{D} = \varepsilon \mathbf{E}, \tag{5.5}$$

$$\mathbf{B} = \mu \mathbf{H}, \tag{5.6}$$

$$\mathbf{F} = q(\mathbf{E} + \mathbf{v} \times \mathbf{B}). \tag{5.7}$$

Following our custom of deriving further equations from our basic set, let us take the curl of eqn (5.2):

$$\nabla \times \nabla \times \mathbf{E} = -\frac{\partial}{\partial t}(\nabla \times \mathbf{B}). \tag{5.8}$$

Using further the vector identity (A.6) and eqns (5.1) and (5.4) to (5.6) we get

$$\nabla^2 \mathbf{E} - \nabla\nabla \cdot \mathbf{E} - \mu\varepsilon \frac{\partial^2 \mathbf{E}}{\partial t^2} = \mu \frac{\partial \mathbf{J}}{\partial t}. \tag{5.9}$$

A similar equation may be derived for the magnetic field by taking the curl of eqn (5.1), yielding

$$\nabla^2 \mathbf{H} - \mu\varepsilon \frac{\partial^2 \mathbf{H}}{\partial t^2} = -\nabla \times \mathbf{J}. \tag{5.10}$$

† The real reason for writing them down again and again is quite different. It is a well known tenet of psychology that the more often you see a statement in print the more likely you are to accept its truth. Mind you, there are exceptions to this rule.

It is true again that in certain cases the scalar and vector potentials may be preferable to the electric and magnetic field quantities so we do the derivations in terms of **A** and ϕ as well. Remember that

$$\mathbf{B}=\nabla\times\mathbf{A} \tag{3.8}$$

and

$$\mathbf{E}=-\frac{\partial\mathbf{A}}{\partial t}-\nabla\phi, \tag{4.13}$$

which substituted into eqn (5.1) yields

$$\nabla\nabla\cdot\mathbf{A}-\nabla^2\mathbf{A}=\mu\mathbf{J}-\varepsilon\mu\frac{\partial^2\mathbf{A}}{\partial t^2}-\varepsilon\mu\,\nabla\frac{\partial\phi}{\partial t}. \tag{5.11}$$

Remember that we are free to choose the divergence of **A**. By choosing

$$\nabla\cdot\mathbf{A}=-\mu\,\varepsilon\frac{\partial\phi}{\partial t}, \tag{5.12}$$

eqn (5.11) will simplify to

$$\nabla^2\mathbf{A}-\varepsilon\,\mu\frac{\partial^2\mathbf{A}}{\partial t^2}=-\mu\mathbf{J}. \tag{5.13}$$

If you compare this with eqn (3.11) (p. 57) you will find that we have now a new term, the second derivative with respect to time.

Substituting now eqn (4.13) (p. 89) into (5.3) we get

$$-\nabla^2\phi-\frac{\partial}{\partial t}\nabla\cdot\mathbf{A}=\frac{\rho}{\varepsilon}, \tag{5.14}$$

which in view of eqn (5.12) reduces to

$$\nabla^2\phi-\varepsilon\mu\frac{\partial^2\phi}{\partial t^2}=-\frac{\rho}{\varepsilon}. \tag{5.15}$$

Again, the second derivative term is added to the static equation.

5.3. The displacement current in a capacitor

We have discussed the parallel-plate capacitor at quite a length in Chapter 2. We shall now extend the discussion to the currents flowing onto and away from the plates.

Let us first look at the process of charging up a capacitor. When switch A in Fig. 5.1(a) is closed a current flows from the battery into the plates as shown in Fig. 5.1(b). The electron current is of course opposite so that negative charges leave the upper plate and accumulate on the lower plate. The charging process comes to an end when the voltage across the capacitor becomes identical with the battery voltage.

What about discharging a capacitor? Then the direction of current flow is reversed (Fig. 5.1(c)), electrons desert the lower plate and enter the upper plate. All this makes good sense. The electrons are forbidden to move across the gap between the plates, so they have to use the good services of their fellow electrons in the connecting wires to get across the long way. While the capacitor is being charged up all the electrons in the circuit move in one direction, and all the electrons move in the other direction under discharging.

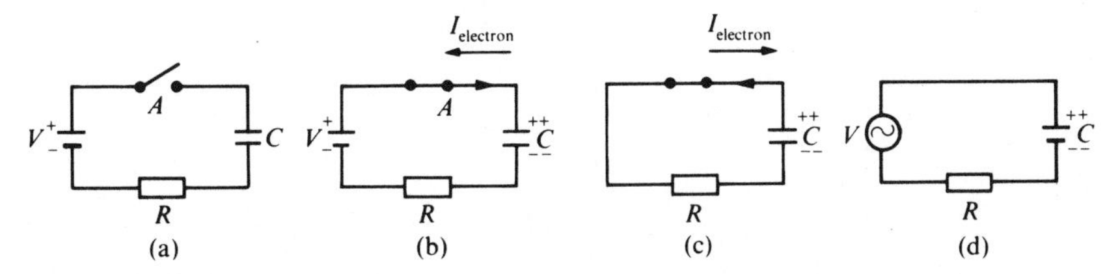

FIG. 5.1. An electric circuit illustrating the charging and discharging of a capacitor.

If we connected an a.c. source as shown in Fig. 5.1(d) we coud draw the same conclusions: charges move one way and then the other way when the polarity of the source changes. There is no current all the way round because there is no motion of electrons between the plates. All this is in perfect agreement with common sense whether we think of the flow of particles or of a fluid but very much against the concepts of circuit theory. If you think of a capacitor as a circuit element you must allow currents to flow through the capacitor.

'Wouldn't life be easier', you could comment at this stage, 'if some other type of current would flow between the plates. Then circuits containing capacitors would not need to be distinguished. We could just say then that such and such a current is flowing *through* the capacitor.'

Well, let us make the effort; calculate the particle current in the wires and then postulate that the same current must flow between the plates.

From the definition of capacitance we know that

$$q = CV. \tag{5.16}$$

Remember further that for a parallel-plate capacitor

$$V = Ed \tag{5.17}$$

and

$$C = \varepsilon \frac{S}{d}, \tag{5.18}$$

where S is the area of the plates and d is the distance between them. Hence the current is

$$I = \frac{\partial q}{\partial t} = C\frac{\partial V}{\partial t} = \varepsilon S\frac{\partial E}{\partial t}. \tag{5.19}$$

So in order to save our picture that an electric current must go all the way round, we could postulate that the *same* current flows between the capacitor plates. Calling that current the displacement current we get

$$I_d = \varepsilon S\frac{\partial E}{\partial t} = S\frac{\partial D}{\partial t}, \tag{5.20}$$

and the displacement current density is

$$\mathbf{J}_d = \frac{\partial \mathbf{D}}{\partial t}. \tag{5.21}$$

Now what do Maxwell's equations say on the subject? Taking the divergence of eqn (5.1) we find that

$$\nabla \cdot \left(\mathbf{J} + \frac{\partial \mathbf{D}}{\partial t}\right) = 0. \tag{5.22}$$

There appears to be a quantity (often called *total current density*)

$$\mathbf{J}_t = \mathbf{J} + \frac{\partial \mathbf{D}}{\partial t} \tag{5.23}$$

which is conserved.

For the capacitor $\mathbf{J} = 0$ between the plates so the total current is carried by $\partial \mathbf{D}/\partial t$. But Maxwell's equations say a lot more than that. In any geometry, under any conditions it is the sum of the two currents that is conserved.

5.4. The continuity equation

The continuity equation is a mathematical expression of the principle of mass conservation. It comes into the theory of all fluids. Since in this book our interest is concentrated upon the behaviour of charged fluids, we shall rather be concerned with the principle of charge conservation.†

Let us take a volume τ bounded by the closed surface S. The total amount of charge in τ is

$$q = \int_\tau \rho(t)\, d\tau \tag{5.24}$$

at time t. The net current (current out minus current in) flowing through the

† If there is only one kind of mobile particle present then charge conservation is equivalent to mass conservation.

closed surface S is

$$\int_S \mathbf{J} \,.\, \mathrm{d}\mathbf{S}. \tag{5.25}$$

Hence the net amount of charge flowing out of the volume in a time interval Δt is

$$q_{\mathrm{net}} = \Delta t \int \mathbf{J} \,.\, \mathrm{d}\mathbf{S}, \tag{5.26}$$

which must be equal to the change in the amount of enclosed charge during the same interval Δt coming to

$$\int_\tau \{\rho(t) - \rho(t+\Delta t)\}\, \mathrm{d}\tau \cong -\Delta t \int_\tau \frac{\partial \rho}{\partial t}\, \mathrm{d}\tau. \tag{5.27}$$

Since eqns (5.26) and (5.27) express the same thing we get the equation

$$-\int_\tau \frac{\partial \rho}{\partial t}\, \mathrm{d}\tau = \int_S \mathbf{J} \,.\, \mathrm{d}\mathbf{S}. \tag{5.28}$$

Applying Gauss's theorem to the right-hand side we end up with

$$\frac{\partial \rho}{\partial t} + \nabla \,.\, \mathbf{J} = 0, \tag{5.29}$$

which is known as the *continuity equation.*

If Maxwell's equations lay any claim to general validity they should reproduce eqn (5.29), and reproduce they will. We only need to substitute eqn (5.3) into eqn (5.22) and we get the required result.

Note that the presence of the $\partial \mathbf{D}/\partial t$ term in eqn (5.3) is absolutely essential. Without it charge is not conserved.

Now imagine yourself in the following situation. In your opinion charge should be conserved but the equation in front of you, supported by an impressive volume of experimental evidence, looks like

$$\nabla \times \mathbf{H} = \mathbf{J}, \tag{5.30}$$

which does not lead to charge conservation. What will you do? Add the term $\partial \mathbf{D}/\partial t$. This is what James Clerk Maxwell did in the seventh decade of the last century.

Let us now return to a somewhat less important and more recent historical event, to our discussion of charge rearrangement inside a conductor. I said, 'if we wait patiently for a few picoseconds until the charges rearrange themselves . . .' We are now in a position to confirm or disprove that statement. Assuming finite conductivity and the $\mathbf{J} = \sigma \mathbf{E}$ relationship it follows that

$$\nabla \,.\, \mathbf{J} = \sigma \nabla \,.\, \mathbf{E} = \frac{\sigma}{\varepsilon} \rho, \tag{5.31}$$

which substituted into eqn (5.29) yields the differential equation

$$\frac{\partial \rho}{\partial t}+\frac{\sigma \rho}{\varepsilon}=0. \tag{5.32}$$

The solution is

$$\rho=\rho_0\,\mathrm{e}^{-(\sigma/\varepsilon)t}. \tag{5.33}$$

Hence any excess charge density in the interior of a conductor will disappear with a time constant $T=\varepsilon/\sigma$. For a good conductor like copper $\sigma \cong 5{\cdot}7\times 10^7\ \mathrm{S\,m^{-1}}$ and $T=1{\cdot}5\times 10^{-19}$ s, a pretty short time. Thus when I pleaded for a few picoseconds, I asked for your patience for an unnecessarily long time.

5.5. Poynting's theorem

Let us do first a little mathematics: multiply eqn (5.1) by **E**, eqn (5.2) by **H** and subtract the former expression from the latter one. We get

$$\mathbf{H}\,.\,(\nabla\times\mathbf{E})-\mathbf{E}\,.\,(\nabla\times\mathbf{H})=-\mathbf{H}\,.\,\frac{\partial \mathbf{B}}{\partial t}-\mathbf{E}\,.\,\frac{\partial \mathbf{D}}{\partial t}-\mathbf{E}\,.\,\mathbf{J}. \tag{5.34}$$

Using the vector identity eqn (A.4) we may rewrite eqn (5.34) in the form

$$\nabla\,.\,(\mathbf{E}\times\mathbf{H})=-\mathbf{H}\,.\,\frac{\partial \mathbf{B}}{\partial t}-\mathbf{E}\,.\,\frac{\partial \mathbf{D}}{\partial t}-\mathbf{E}\,.\,\mathbf{J}. \tag{5.35}$$

Let us analyse this equation term by term. What is **E . J**? That is just the rate at which work is done per unit volume as may be seen by the following derivation

$$\frac{\rho\mathbf{F}\,.\,\mathbf{v}}{q}=\rho\mathbf{v}\,.\,(\mathbf{E}+\mathbf{v}\times\mathbf{B})=\rho\mathbf{v}\,.\,\mathbf{E}=\mathbf{J}\,.\,\mathbf{E}. \tag{5.36}$$

Next we shall look at the expression

$$\mathbf{E}\,.\,\frac{\partial \mathbf{D}}{\partial t}+\mathbf{H}\,.\,\frac{\partial \mathbf{B}}{\partial t}=\frac{\partial}{\partial t}(\tfrac{1}{2}\varepsilon E^2+\tfrac{1}{2}\mu H^2)=\frac{\partial W_{\mathrm{d}}}{\partial t}, \tag{5.37}$$

where W_{d} is defined by the above equation. The first term in the bracket was derived in Section 2.12 for the energy density of the electrostatic field. The second could also have been derived (though we did not perform the calculations) as the formula for the energy density in a magnetostatic field.

What about **E** × **H**? Before answering that question I would like you to have another look at eqn (5.35). Introducing the notation

$$\mathbf{P}_{\mathrm{d}}=\mathbf{E}\times\mathbf{H}, \tag{5.38}$$

eqn (5.35) takes the form

$$-\frac{\partial W_d}{\partial t}=\nabla\,.\,\mathbf{P}_d+\mathbf{E}\,.\,\mathbf{J}. \tag{5.39}$$

Our interpretation of the vector $\mathbf{P}_d$ (called the Poynting vector) may now proceed on the same lines as for the continuity equation in Section 5.4. Remember

$$-\frac{\partial \rho}{\partial t}=\nabla\,.\,\mathbf{J} \tag{5.40}$$

meant that a change in charge density must be accompanied by a net outflow of current. Then eqn (5.39) must mean that if the energy density changes inside a volume τ, it is either destroyed (i.e. transformed into mechanical energy or heat) or it flows out of the volume. Thus besides the local conservation of charge, we have an equation for the local conservation of energy.† But there is an essential difference between the two cases. When discussing charge conservation we *started* with the concepts of charge and current and *then* derived the continuity equation. For energy conservation our approach was much more dubious. First we derived the equation and then found an interpretation. Is W_d really the energy density? Well, $\frac{1}{2}\varepsilon E^2$ was the static energy density, so one might be prepared to accept that the same formula will apply for time varying electric fields as well. Does $\mathbf{P}_d$ really give the direction and magnitude of energy flow? We can't be sure. For one thing we could always add a vector $\nabla\times\mathbf{M}$ to $\mathbf{P}_d$ without affecting eqn (5.39) but even worse, by adding the equation

$$-\frac{\partial G}{\partial t}=\nabla\,.\,\mathbf{T} \tag{5.41}$$

to eqn (5.39) we get a new equation of the form

$$-\frac{\partial}{\partial t}(W_d+G)=\nabla\,.\,(\mathbf{P}_d+\mathbf{T})+\mathbf{E}\,.\,\mathbf{J}. \tag{5.42}$$

Are we now to interpret W_d+G as energy density and $\mathbf{P}_d+\mathbf{T}$ as energy flow? We appear to be in a bit of trouble. If in trouble with the theory go back to experiments. True, but unfortunately energy density and energy flow at a given point in space are not directly measurable. So are we completely stuck? No, there is an escape route offered by the theory of relativity. As you know, mass and energy are equivalent hence any energy density located somewhere in space should be identifiable by its gravitational effect. I

† Note that local conservation of energy means more than just conservation of energy. According to the latter the total amount of energy in the universe cannot change. The former one says in addition that energy cannot just appear out of the blue. If the energy increases at point B at the expense of the energy at point A then there must be a flow of energy from A to B.

believe this experiment has not been performed as yet but apparently our definition of energy density and energy flow fits excellently into other parts of relativity theory so you should have no qualms about accepting them.

What is eqn (5.39) good for? Both for working out the total stored energy and for determining the direction and magnitude of energy flow. Interestingly, the latter application brings forward a number of unexpected results.

We shall look at only one example (if you want to gain more experience solve Examples 5.1 and 5.5), at the energy flow along a d.c. transmission line (Fig. 5.2(a)).†

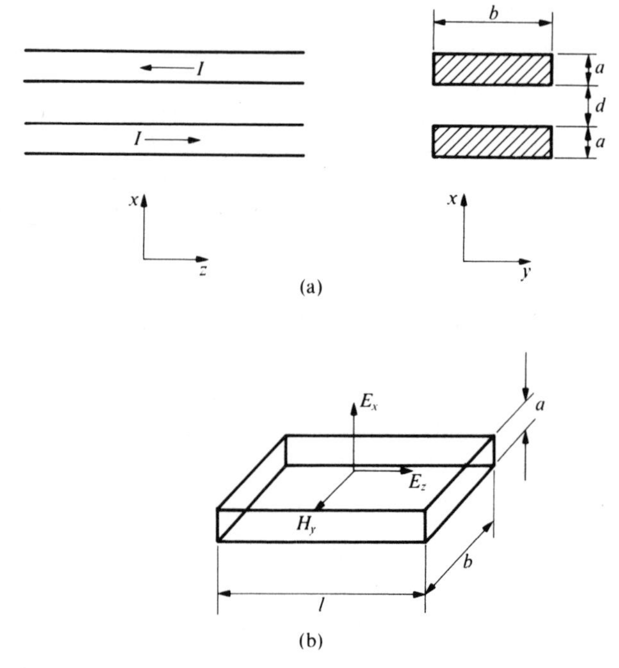

FIG. 5.2. (a) A d.c. transmission line carrying equal and opposite currents. (b) The electric and magnetic fields at the top of the lower conductor.

The transverse dimensions of the two conductors are deliberately chosen in such a way that the calculation of the fields should be easy. We may then regard both the electric and magnetic fields constant between the conductors and zero outside, leading to the equations

$$\mathbf{E}=E_x\mathbf{i}_x=\frac{V}{d}\mathbf{i}_x \quad \text{and} \quad \mathbf{H}=H_y\mathbf{i}_y=\frac{I}{b}\mathbf{i}_y. \tag{5.43}$$

† You could rightfully ask why a d.c. problem is presented among fast-varying phenomena. The answer is that the inclusion of the displacement current term was essential for arriving at the physical picture of power flow. But having done so, the simplest illustration is a d.c. transmission line.

Hence the Poynting vector may be obtained as

$$\mathbf{P}_d = \frac{VI}{bd}\mathbf{i}_z. \tag{5.44}$$

It may be seen from the above equation that the energy flows along the transmission line, a satisfying result (if energy flowed in any other direction we would be in trouble). How large is the total power carried? We may obtain that by integrating the Poynting vector over the space between the conductors, as follows

$$P = \int_S \mathbf{P}_d \cdot d\mathbf{S} = \int_0^d \int_0^b \frac{VI}{bd}\,dx\,dy = VI, \tag{5.45}$$

another trivial result.† But you will get the promised shock if you give another thought to the actual location of the Poynting vector. It is located in the space in between the conductors. The inevitable conclusion is that the energy transmitted by the line is carried in the open space and *not* in the conductors. Are the conductors necessary at all? Before answering this question let us see what happens inside the conductor. Owing to the finite conductivity of the material there will be a longitudinal electric field

$$E_z = \frac{J}{\sigma} = \frac{I}{ab\sigma}. \tag{5.46}$$

The non-zero components of the electric and magnetic field vectors on the top surface of the lower conductor are shown in Fig. 5.2(b). We still have the energy flow in the longitudinal direction due to the $E_x H_y$ term but there is also a new component of the Poynting vector

$$P_{dx} = -E_z H_y \tag{5.47}$$

pointing straight into the conductor. By integrating over the surface of the conductor for a length l of the transmission line we get for the total power entering

$$P_e = \int_0^l \int_0^b E_z H_y\,dz\,dy = \frac{I}{ab\sigma}\frac{I}{b}lb = I^2 R, \tag{5.48}$$

where

$$R = \frac{l}{ab\sigma} \tag{5.49}$$

is the resistance of the material.

Now we are getting something really interesting. The conductors are not innocent bystanders; they are responsible for the absorption of a lot of useful

† Note that the dimension of $\mathbf{P}_d$ is energy per unit time per unit surface. Integrating over a surface yields the dimension of power.

energy which would otherwise reach the consumers. Hence the presence of the conductors is not only unnecessary but positively harmful.

The foregoing argument is attractive, but something must be wrong with it. If we dispense with the conductors how would the energy know which way to flow? We must have the conductors in order to give a preferred direction. Mathematically speaking, the conductors are there to establish the boundary conditions. So they perform a useful service but we have to pay for that by permitting the deflection of a certain part of the energy into the conductors. The amount of energy moving into the conductors per unit time is just equal to I^2R for each conductor as you would calculate from circuit theory.

Does all this mean that the circuit viewpoint is wrong? Yes, as far as the propagation of energy is concerned it is quite wrong. 'But', you may ask, 'does it make any difference if I carry on thinking in terms of currents and voltages when envisaging transfer of energy? The approach is simpler and after all it has never been contradicted by experience.'

I believe you will be better off by accepting the field picture because it is a more general one. By thinking of the Poynting vector you will loosen the straight-jacket imposed upon your imagination by circuit theory. You may, for example, strike upon the following idea. If the only role of the conductors is to guide the energy along the line then surely we don't need two of them. Or you might ask: what is so specific about conductors, why are they singled out for guiding the energy along? Couldn't we use dielectrics instead? Both questions deserve further investigation, and we shall return to them some time later.

By the way, I forgot to mention, eqn (5.39) together with its interpretation is known as Poynting's theorem, derived by Poynting in 1884.

5.6. Boundary conditions

We have derived the boundary conditions for the electrostatic case in Section 2.9 and simply stated them for the magnetostatic case in Section 3.9. How do they modify when time dependence is taken into account? They remain the same.

I don't intend to repeat the whole performance again. They all follow the same type of reasoning so it will suffice to do the derivation for only one of them, namely the tangential component of the electric field.

First I want to remind you of eqn (4.10)

$$\oint \mathbf{E} \cdot \mathrm{d}\mathbf{s} = -\frac{\partial \Phi}{\partial t} \tag{4.10}$$

and choose the path of the line integral at the boundary of the two media as shown in Fig. 2.38 (p. 45). We get

$$\oint \mathbf{E} \cdot \mathrm{d}\mathbf{s} = (E_{t2} - E_{t1})\, \mathrm{d}s. \tag{5.50}$$

As $\mathrm{d}l \to 0$ the area of the closed loop goes to zero as well, and the magnetic flux through it vanishes. Thus the right-hand side of eqn (4.10) is zero in the limit, leading to

$$E_{t1} = E_{t2}, \tag{5.51}$$

just as in the static case. In fact all other boundary conditions remain unchanged. In order to have them collected in one place and to show them in vectorial form, I write them out again:

$$\begin{aligned} \mathbf{n} \times (\mathbf{E}_2 - \mathbf{E}_1) &= 0, \qquad & \mathbf{n} \cdot (\mathbf{D}_2 - \mathbf{D}_1) &= \rho_s, \\ \mathbf{n} \times (\mathbf{H}_2 - \mathbf{H}_1) &= \mathbf{K}, \qquad & \mathbf{n} \cdot (\mathbf{B}_2 - \mathbf{B}_1) &= 0. \end{aligned} \tag{5.52}$$

where $\mathbf{n}$ is the surface normal.

In many forthcoming problems we shall use the approximation that one of the media is a perfect conductor. Since the electric field must be zero inside a perfect conductor, the boundary condition for the electric field and electric flux density reduces to

$$\mathbf{n} \times \mathbf{E}_1 = 0 \quad \text{and} \quad \mathbf{n} \cdot \mathbf{D}_1 = \rho_s. \tag{5.53}$$

Time-varying magnetic fields will also be excluded from the interior of a perfect conductor because the time varying flux induces non-decaying currents which set up a magnetic field of opposite sign.† Hence the boundary conditions take the form

$$\mathbf{n} \times \mathbf{H}_1 = \mathbf{K} \quad \text{and} \quad \mathbf{n} \cdot \mathbf{B}_1 = 0. \tag{5.54}$$

5.7. Plane waves

We shall now investigate the simplest possible form of electromagnetic waves. The solution may be obtained from our basic equations with the aid of the following assumptions:

1. the whole space is void of free charges;
2. no currents flow;
3. the electric field has only one single component in the x direction;
4. there is no variation in the transverse plane, $\partial/\partial x = \partial/\partial y = 0$.

In view of the above assumptions, eqn (5.9), the differential equation for the electric field, simplifies to

$$\frac{\partial^2 E_x}{\partial z^2} - \frac{1}{c_m^2}\frac{\partial^2 E}{\partial t^2} = 0, \tag{5.55}$$

† Note that a perfect conductor will not expel a constant magnetic field unless it is a perfect diamagnet as well. In the present chapter where we are concerned with fast-varying phenomena this need not concern us.

where

$$c_m^2 = 1/\varepsilon\mu. \tag{5.56}$$

The general solution of the above equation is

$$E_x = C_1 f(z - c_m t) + C_2 f(z + c_m t), \tag{5.57}$$

which can be easily proved by substituting eqn (5.57) into (5.55).

What sort of function is f? Any reasonable-looking function will do. Take for example $f(z)$ at $t = 0$, as the function shown in Fig. 5.3 by a solid line. It has a certain finite value in the interval $z_1 < z < z_2$ and zero outside. How will

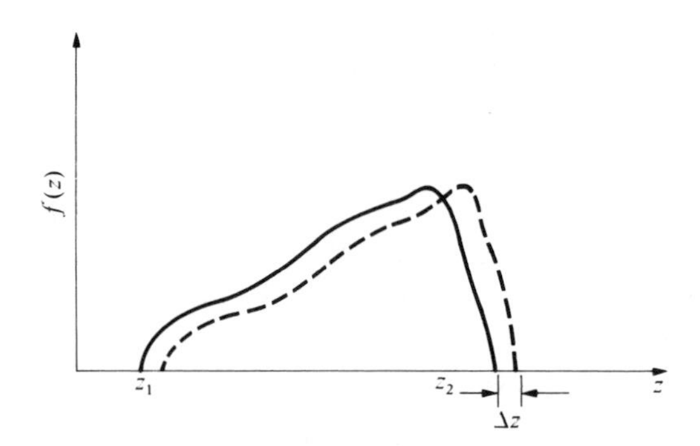

FIG. 5.3. A travelling waveform.

the same function look a time Δt later? It will be shifted in the direction of the positive z axis by the amount $c_m \Delta t$, as shown by dotted lines in the same figure.

The physical picture is clear: during the time interval Δt the electric field intensity advanced by the distance

$$\Delta z = c_m \Delta t, \tag{5.58}$$

i.e. it has moved with a velocity c_m (by similar arguments the second term of eqn (5.57) yields a wave moving backwards). In a vacuum

$$c_m = \frac{1}{\sqrt{(\varepsilon_0\mu_0)}} = 2{\cdot}998 \times 10^8 \text{ m s}^{-1} = c, \tag{5.59}$$

the velocity of light. This was a result that made the blood run faster in the veins of the physicists of the 1870s and 1880s. That light is an electromagnetic wave is now taught in secondary schools and, I would claim, it has become part of general (and hence undigested) knowledge. The other day I met an arts graduate who knew about it.

Having got the solution for the electric field, the corresponding magnetic field may be obtained from eqn (5.2). The left-hand side takes the form

$$\nabla \times \mathbf{E} = \begin{vmatrix} \mathbf{i}_x & \mathbf{i}_y & \mathbf{i}_z \\ 0 & 0 & \dfrac{\partial}{\partial z} \\ E_x & 0 & 0 \end{vmatrix} = \frac{\partial E_x}{\partial z}\mathbf{i}_y, \tag{5.60}$$

leading to the differential equation for H_y

$$-\mu \frac{\partial H_y}{\partial t} = \frac{\partial E_x}{\partial z}. \tag{5.61}$$

Noting that

$$\frac{\partial}{\partial z} f(z \pm c_{\mathrm{m}} t) = f'(z \pm c_{\mathrm{m}} t) \tag{5.62}$$

and

$$\frac{\partial}{\partial t} f(z \pm c_{\mathrm{m}} t) = \pm c_{\mathrm{m}} f'(z \pm c_{\mathrm{m}} t), \tag{5.63}$$

the solution of the above differential equation is

$$H_y = \frac{1}{\mu c_{\mathrm{m}}} \{C_1 f(z - c_{\mathrm{m}} t) - C_2 f(z + c_{\mathrm{m}} t)\}. \tag{5.64}$$

Assuming that $C_2 = 0$ (i.e. only a forward-travelling wave is present) the ratio of electric to magnetic field is

$$\frac{E_x}{H_y} = \mu c_{\mathrm{m}} = \sqrt{\left(\frac{\mu}{\varepsilon}\right)} = Z. \tag{5.65}$$

For a vacuum

$$\frac{E_x}{H_y} = \sqrt{\left(\frac{\mu_0}{\varepsilon_0}\right)} = Z_0 \cong 120\pi \text{ ohms.} \tag{5.66}$$

is called the impedance of free-space.

This might be a convenient point for summarizing what we have done in this section. First, we chose an open space. Secondly, we excluded variation in the transverse plane so reducing the problem to a one-dimensional one. In the third place we assumed a certain component for the electric field and found that the corresponding magnetic field vector is in the transverse plane perpendicular to the electric field vector. We found further that both vectors travel with the velocity of light in a vacuum and their ratio is constant.

What else would we like to know? The direction in which energy flows. But surely that is in the z direction. The energy must flow in the direction the wave travels. Yes, this is what our physical intuition tells us but it should be checked. In the present case the Poynting vector is given by

$$\mathbf{P}_{\mathrm{d}} = E_x H_y \mathbf{i}_z \tag{5.67}$$

so our physical intuition was right. The actual amount of power carried depends of course on the f function, i.e. on the shape of the travelling waveform. We shall return to that problem in Section 5.9, where sinusoidal waveforms will be studied.

5.8. The single-frequency assumption and the exponential time dependence

I shall introduce here a significant simplification that will make our life easier for the rest of this course. I shall assume that the time variation is imposed on the system, that everything varies as sin ωt.

Is this a reasonable assumption? One of the best, I would say. We shall be able to discuss practically all phenomena of interest to engineers by restricting the investigation to harmonic excitation. One of the reasons why we can do that is that we are concerned with fast-varying phenomena, hence transients take a negligibly short time. Another reason is that most waves of interest are man-made and designed to occupy a small frequency band. A third reason is that by doing all our calculations for single-frequency waves does not mean that the validity of the results is restricted to a single frequency. We can have our cake and eat it. We save a lot of mathematical effort by assuming a single frequency, but at the end of the day when all the calculations are done we are still at liberty to choose the actual numerical value of ω to be substituted into the equations. Of course some of the 'constants' might turn out to be dependent on frequency; a good example is the relative permittivity. But if $\varepsilon(\omega)$ is known we use its value for the particular frequency of interest and everything is all right.

I said sinusoidal time dependence, but should really have said exponential time dependence. It is an old trick (used liberally not only in circuit theory but in many other branches of engineering as well) to assume the time dependence in the form

$$f(t) = \mathrm{e}^{\mathrm{j}\omega t}, \tag{5.68}$$

and do all the mathematics in terms of the above function. This is perfectly legitimate provided all the equations are linear.† We may then regard the functions

$$\operatorname{Re} f(t) \quad \text{or} \quad \operatorname{Im} f(t) \tag{5.69}$$

(the choice is ours) as expressing the actual time variation.

† And linear they must be, a point worth remembering when you are confronted with real problems in real life. I have seen mistakes in reputable journals caused by the retention of exponential notation in problems not strictly linear.

The main advantage of the choice of exponential time dependence is that from now on

$$\frac{\partial}{\partial t} = \mathrm{j}\omega, \tag{5.70}$$

i.e. differentiation with respect to time is replaced by multiplication by $\mathrm{j}\omega$.

There is no point in rewriting all our equations in Section 5.2 with the aid of the new assumption but just to give you a chance to get used to the new appearance of the equations I shall present here eqns (5.1) and (5.2):

$$\nabla \times \mathbf{H} = \mathbf{J} + \mathrm{j}\omega\varepsilon\mathbf{E} \tag{5.71}$$

and

$$\nabla \times \mathbf{E} = -\mathrm{j}\omega\mu\mathbf{H}. \tag{5.72}$$

One last point I want to make here is that in the problems arising we shall be able to regard ε and μ as constants (they might be dependent on frequency but not on field amplitude) hence it is preferable to use only two of the four variables **E**, **D**, **H**, and **B**. Most physicists prefer to use **B** (and they are right too, **B** is a more fundamental quantity than **H**) but for engineers the use of **H** appears to be more natural. If you look into the literature you will find that the overwhelming majority of learned articles are written in terms of **H**, so we shall follow that custom.

5.9. Plane waves revisited

Armed with our exponential time dependence we shall now re-examine the propagation of plane waves. We shall assume the solution of eqn (5.55) in the form

$$E_x = G(z)\,\mathrm{e}^{\mathrm{j}\omega t}, \tag{5.73}$$

which leads to the differential equation

$$\frac{\partial^2 G(z)}{\partial z^2} + \beta_\mathrm{m}^2 G(z) = 0; \qquad \beta_\mathrm{m} = \omega/c_\mathrm{m}. \tag{5.74}$$

The solution is of course in terms of harmonic functions. Having been sold on the advantages of the exponential form it seems sensible to choose the same form for space dependence, yielding

$$G(z) = G_\mathrm{f}\exp(-\mathrm{j}\beta_\mathrm{m}z) + G_\mathrm{r}\exp(\mathrm{j}\beta_\mathrm{m}z), \tag{5.75}$$

where G_f and G_r are constants (generally complex) and β_m is the propagation coefficient. The period in space is what we call the wavelength; it comes to

$$\lambda_\mathrm{m} = \frac{2\pi}{\beta} = \frac{c_\mathrm{m}}{f}. \tag{5.76}$$

In free space, $\lambda = c/f$ and $\beta = \omega/c = 2\pi/\lambda = \beta_0$.

If we include time variation as well, the complete solution for the electric field in free space is of the form

$$E_x = G_f \exp\{j(\omega t - \beta_0 z)\} + G_r \exp\{j(\omega t + \beta_0 z)\}. \tag{5.77}$$

The differential equation for H_y (eqn (5.61)) may be solved in a similar manner, leading to

$$H_y = \frac{1}{\mu_0 c}[G_f \exp\{j(\omega t - \beta_0 z)\} - G_r \exp\{j(\omega t + \beta_0 z)\}]. \tag{5.78}$$

We could, in fact, have immediately written eqns (5.77) and (5.78) from eqns (5.57) and (5.64) by taking

$$f(z \mp ct) = \exp\{j\beta_0(ct \pm z)\}. \tag{5.79}$$

Whichever way we look at it, we have two waves, one travelling forwards the other one backwards.

Having obtained E_x and H_y in terms of known functions of z and t we could now determine the average power passing through a unit surface.

Let's pause here for a moment. Our solution in terms of $\exp j\omega t$ is perfectly legitimate because eqn (5.55) is linear. But in order to form a quadratic expression we have to return to real functions as illustrated in the following example. Let's take two complex quantities of the form

$$A = a \exp j\varphi_1 \exp j\omega t \quad \text{and} \quad B = b \exp j\varphi_2 \exp j\omega t, \tag{5.80}$$

where a, b, φ_1, φ_2 are real. According to the recipe of eqn (5.69) we shall take the real parts as the physical quantities. The time average (average for a period of oscillation) is then

$$\overline{AB} = \frac{1}{2\pi}\int_0^{2\pi} (\text{Re } A)(\text{Re } B)\, d(\omega t) = \frac{ab}{2}\cos(\varphi_1 - \varphi_2). \tag{5.81}$$

You may see now that we can get the above expression from eqn (5.80) by forming

$$\overline{AB} = \tfrac{1}{2}\,\text{Re}(AB^*), \tag{5.82}$$

where B^* is the complex conjugate of B.

It is obvious that we have arrived at eqn (5.82) in a purely artificial manner. There is no physics behind it, we just wanted to reproduce the correct formula without abandoning the complex functions close to our hearts.†

Hence, using the complex notation, we can write the time average of the Poynting vector as

$$\mathbf{P}_d = \tfrac{1}{2}\,\text{Re}(\mathbf{E} \times \mathbf{H}^*). \tag{5.83}$$

† The same motives lead to the same kind of formula for the power in an a.c. circuit.

In the present case we need to substitute eqns (5.77) and (5.78) into (5.83), yielding

$$P_d = \frac{1}{2Z_0}(|G_f|^2 - |G_r|^2). \tag{5.84}$$

Quite obviously, the net power in the positive z direction is equal to the power carried by the forward travelling wave minus the power carried by the backward travelling wave.

I haven't finished with plane waves yet; I have to say a few words about the generality of the solution. Did we restrict generality to such an extent that the resulting mathematical solution bears little relationship to practical problems? No, not at all. A space free of charges and currents is not a bad description for the media in which light rays or radio waves travel. What about the one-dimensional assumption? Surely, no wavefronts extend to infinity. Quite true, but if the source of the electromagnetic waves (the sun or a radio transmitter) is far away, the local wavefront may be regarded a plane one with very good approximation, so we are all right again.

The assumption that will not bear closer scrutiny is that the electric field has only an x component. Why an x component? Naturally, other components have the same right to existence. We could have, for example, chosen an E_y component and would have found the magnetic field pointing in the $-x$ direction if the wave is travelling in the $+z$ direction. It turns out that for a plane wave the electric and magnetic vectors are always in a plane perpendicular to the direction of propagation, and in addition they are perpendicular to each other. This is why they are called TEM (short for *transverse electromagnetic*) *waves.*

What happens if we superimpose two such waves? Assume, for example, that two plane waves travel in the z direction whose electric fields are given by the equations

$$\mathbf{E}_1 = E_x \mathbf{i}_x = G_f \exp\{\mathrm{j}(\omega t - \beta_0 z)\}\mathbf{i}_x \tag{5.85}$$

and

$$\mathbf{E}_2 = E_y \mathbf{i}_y = -\mathrm{j} G_f \exp\{\mathrm{j}(\omega t - \beta_0 z)\}\mathbf{i}_y, \tag{5.86}$$

and let us investigate the resultant electric field as a function of time at a fixed point in space, say at $z = 0$.

You have to remember that it is the real part that counts physically so that

$$\mathbf{E}_1 = G_f \cos \omega t \mathbf{i}_x \tag{5.87}$$

and

$$\mathbf{E}_2 = G_f \sin \omega t \mathbf{i}_y, \tag{5.88}$$

where for simplicity G_f was taken real. Quite clearly, the resultant vector $\mathbf{E} = \mathbf{E}_1 + \mathbf{E}_2$ rotates in the x, y plane with an angular velocity ω. This is just to show you that there is no need for the electric field vector always to remain in the same plane. If it does, we talk of linear polarization. If it rotates as in our previous example, we talk of circular polarization. Naturally the two components need not be of equal magnitude and might not have a phase difference of exactly 90°. In that case as you may have guessed the tip of the resultant vector describes an ellipse (see Example 5.3).

5.10. Reflection by a perfect conductor

Let us start again with our plane wave which has E_x and H_y components and travels in the $+z$ direction. We shall now place a perfect conductor into the half infinite space $z > z_0$ (Fig. 5.4(a)). What will happen to the wave?

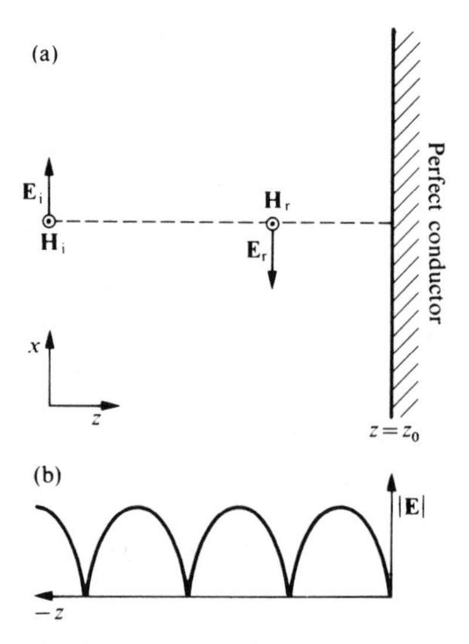

FIG. 5.4. (a) A plane wave incident perpendicularly upon a perfect conductor. (b) Standing-wave pattern.

Well, it cannot penetrate into a perfect conductor, it cannot accumulate in front of the conductor; the only thing it can do is to turn back. Hence the total electric field must be describable as the sum of two waves

$$E_x = G_f \exp(-j\beta_0 z) + G_r \exp(j\beta_0 z). \tag{5.89}$$

As you may notice we have omitted the $\exp j\omega t$ time dependence. It is a factor always present so we don't need to write it out explicitly each time.

The boundary condition to be satisfied is $E_x = 0$ at $z = z_0$, i.e.

$$G_f \exp(-j\beta_0 z_0) + G_r \exp(j\beta_0 z_0) = 0, \tag{5.90}$$

whence

$$G_r = -G_f \exp(-2j\beta_0 z_0). \qquad (5.91)$$

Substituting eqn (5.91) into eqn (5.89) we get the form that now satisfies the boundary condition

$$E_x = G_f\{\exp(-j\beta_0 z) - \exp(-2j\beta_0 z_0)\exp(j\beta_0 z)\}$$
$$= -2jG_f \exp(-j\beta_0 z_0) \sin \beta_0(z - z_0). \qquad (5.92)$$

This is clearly a standing wave (Fig. 5.4(b)).

Next we shall consider a plane wave obliquely incident upon the perfect conductor at $z = z_0$, and polarized in the x, z plane (Fig. 5.5). Physically, this is in no way different from a plane wave propagating in the direction of the z axis; mathematically it looks a bit more complicated. How can we

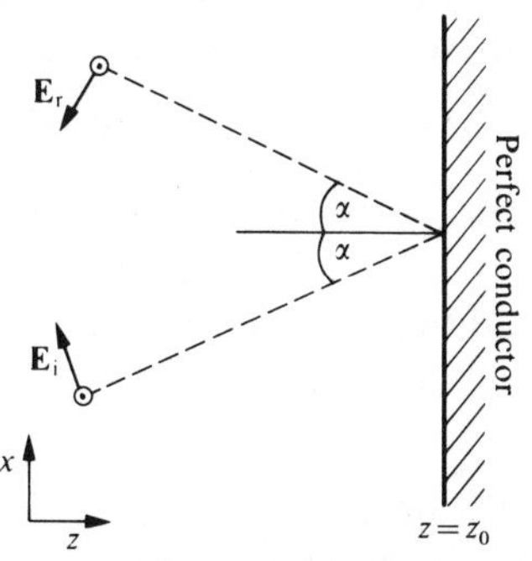

FIG. 5.5. Oblique incidence of plane wave.

find the mathematical form? Well, a plane wave travelling in the $\mathbf{i}_z$ direction has been described as $\exp(-j\beta_0 z)$. It is a plane wave because the constant phase surfaces are the z = constant planes. If we want to describe a plane wave travelling in the $\mathbf{i}_\beta$ direction we must ensure that the phase will be constant in the planes perpendicular to $\mathbf{i}_\beta$. This can be easily achieved by writing $\exp(-j\boldsymbol{\beta}_0 \cdot \mathbf{r})$ where $\boldsymbol{\beta}_0 = \beta_o i_\beta$ and $\bar{r}$ is a radius vector. Then at any point of a plane perpendicular to $\mathbf{i}_\beta$ the scalar product $\boldsymbol{\beta}_0 \cdot \mathbf{r}$ is a constant as may be seen in Fig. 5.6.

The electric field of the incident wave shown in Fig. 5.5 may then be expressed as follows:

$$\mathbf{E}_i = G_f(\cos\alpha\,\mathbf{i}_x - \sin\alpha\,\mathbf{i}_z)\exp\{-j\beta_0(z\cos\alpha + x\sin\alpha)\}. \qquad (5.93)$$

In which direction will the reflected wave propagate? Common sense, ordinary dynamics, and ray optics all tell us that the angle of reflection will be

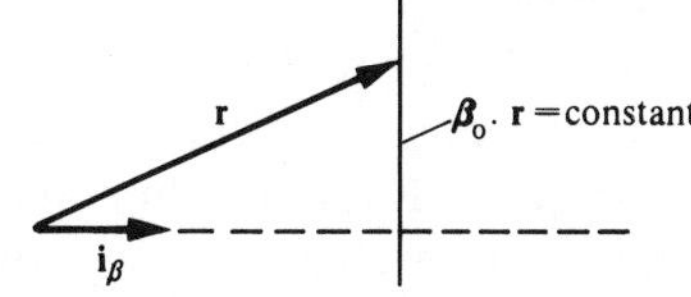

FIG. 5.6. Unit vectors in the direction of propagation.

equal to the angle of incidence. If we know the direction of the reflected wave and the magnetic field vector (it remains unchanged since it is tangential to the surface) $\mathbf{E}_r$ is then defined as shown in Fig. 5.5. Its vectorial form is given as

$$-\cos\alpha\mathbf{i}_x - \sin\alpha\mathbf{i}_z, \tag{5.94}$$

its propagation by

$$\exp(-\mathrm{j}\boldsymbol{\beta}_0\mathbf{r}) = \exp\{-\mathrm{j}\beta_0(x\sin\alpha - z\cos\alpha)\}, \tag{5.95}$$

and its complex amplitude by

$$G_f\,\mathrm{e}^{\mathrm{j}\delta}, \tag{5.96}$$

where δ is a real number showing the relative phase of the incident and reflected waves. Putting these factors together we get

$$\mathbf{E}_r = G_f\,\mathrm{e}^{\mathrm{j}\delta}(-\cos\alpha\mathbf{i}_x - \sin\alpha\mathbf{i}_z)\exp\{-\mathrm{j}\beta_0(x\sin\alpha - z\cos\alpha)\}. \tag{5.97}$$

Our form satisfies energy conservation; the power carried by the incident wave is equal to the power carried by the reflected wave. But we still need to satisfy the boundary condition for the electric field, that the tangential component of $\mathbf{E}_i + \mathbf{E}_r$ must vanish on the boundary. From eqns (5.93) and (5.97)

$$\begin{aligned}(\mathbf{E}_i + \mathbf{E}_r)_t &= -G_f\cos\alpha\exp(-\mathrm{j}\beta_0 x\sin\alpha)\\ &\quad\times[\exp\{\mathrm{j}(\beta_0 z\cos\alpha + \delta)\} - \exp(-\mathrm{j}\beta_0 z\cos\alpha)]\\ &= -2\mathrm{j}G_f\cos\alpha\exp\{-\mathrm{j}(\beta_0 x\sin\alpha + \delta/2)\}\sin\{\beta_0 z\cos\alpha + \delta/2\},\end{aligned} \tag{5.98}$$

which vanishes at $z = z_0$ if

$$\delta = -2\beta_0 z_0\cos\alpha. \tag{5.99}$$

It is worthwhile to note that δ is just the phase corresponding to a wave travelling to and fro from $z = 0$ to $z = z_0$ at an angle α. Had we put our surface in the $z = 0$ plane we need not have bothered about the additional phase. Anyway, with our choice of the position of the conducting surface, eqns (5.93) and (5.97) give the complete solution for the electric field. I am not going to write the expressions for the magnetic fields; they can be easily found.

5.11. Losses and the skin effect

In this section we shall admit that all real media have finite conductivities causing the attenuation of electromagnetic waves.

We may start again with the $\mathbf{J} = \sigma\mathbf{E}$ relationship. Substituting it into eqn (5.9), assuming the absence of free charges and using the $\exp \mathrm{j}\omega t$ time dependence, the differential equation to solve is as follows:

$$\Delta\mathbf{E} + (\omega^2\mu\varepsilon - \mathrm{j}\omega\mu\sigma)\mathbf{E} = 0. \tag{5.100}$$

Resorting again to a one-dimensional picture where the electric field has only an E_x component and propagation is in the z direction, the above differential equation simplifies to

$$\frac{\mathrm{d}^2 E_x}{\mathrm{d}z^2} + (\omega^2\mu\varepsilon - \mathrm{j}\omega\mu\sigma)E_x = 0, \tag{5.101}$$

yielding the solution

$$E_x = G_{\mathrm{f}}\,\mathrm{e}^{-\mathrm{j}\beta z} + G_{\mathrm{r}}\,\mathrm{e}^{\mathrm{j}\beta z}, \tag{5.101(a)}$$

where

$$\beta^2 = \omega^2\mu\varepsilon - \mathrm{j}\omega\mu\sigma. \tag{5.101(b)}$$

For a dielectric it is customary to rewrite eqn (5.101(b)) in the form

$$\beta^2 = \omega^2\mu\varepsilon_0\left(\varepsilon_{\mathrm{r}} - \mathrm{j}\frac{\sigma}{\omega\varepsilon_0}\right) \tag{5.102}$$

and define the term in the bracket as the *complex relative permittivity*

$$\varepsilon' - \mathrm{j}\varepsilon'' = \varepsilon_{\mathrm{r}} - \mathrm{j}\frac{\sigma}{\omega\varepsilon_0}. \tag{5.103}$$

This is just a notation which has no deeper significance but this is the one you will find in handbooks. Conductivity is never quoted for insulators. Instead, ε' and ε'' or ε' and $\tan\delta = \varepsilon''/\varepsilon'$ are given at a few specified frequencies.

For a good dielectric $\varepsilon'' \ll \varepsilon'$, making possible the approximation

$$\beta = \{\omega^2\mu\varepsilon_0(\varepsilon' - \mathrm{j}\varepsilon'')\}^{\frac{1}{2}} \cong (\omega^2\mu\varepsilon_0\varepsilon')^{\frac{1}{2}}\left(1 - \mathrm{j}\frac{\varepsilon''}{2\varepsilon'}\right) = \beta_{\mathrm{m}}(1 - \mathrm{j}\tfrac{1}{2}\tan\delta). \tag{5.104}$$

Accordingly, the forward travelling wave will vary as follows

$$E_x = G_{\mathrm{f}}\exp(-\mathrm{j}\beta_{\mathrm{m}}z)\exp(-\tfrac{1}{2}\beta_{\mathrm{m}}z\tan\delta), \tag{5.105}$$

exhibiting an exponential decay.

The next problem we shall tackle is at the other end of the scale when the material is a good conductor and the inequality $\sigma \gg \omega\varepsilon$ applies. Then

$$\beta \cong (-\mathrm{j}\omega\mu\sigma)^{\frac{1}{2}} = \mp(-1 + \mathrm{j})\zeta; \qquad \zeta = \left(\frac{\omega\mu\sigma}{2}\right)^{\frac{1}{2}}, \tag{5.106}$$

leading to the following form for the forward wave

$$E_x = G_{\mathrm{f}}\,\mathrm{e}^{-\mathrm{j}\zeta z}\,\mathrm{e}^{-\zeta z}. \tag{5.107}$$

The distance at which the field intensity decays to $1/e$ of its value is

$$s = \frac{1}{\zeta} = \left(\frac{2}{\omega\mu\sigma}\right)^{\frac{1}{2}}, \tag{5.107a}$$

called the *skin depth.*

We may now answer the question of what happens when an electromagnetic wave is incident upon a good conductor. The currents flowing on and in the vicinity of the surface will not be able completely to expel the electric field so it will penetrate to a finite distance. The usual measure of penetration is the skin depth s. Let us put in a few figures. Take for example a radio wave at a frequency $f = 500$ kHz and copper for which $\sigma = 5{\cdot}8 \times 10^7$ S m^{-1}, then s turns out to be 0·1 mm, a pretty small distance. The inevitable conclusion is that radio waves won't reach you if you climb into a metal box (so you need a car aerial).

Next we shall consider the problem of current density distribution in a wire of circular cross-section. How is this problem related to the attenuation of plane waves? Well, according to the physical picture developed in Section 5.5 the useful energy is carried in free space and the conductors are only there to account for the losses. In the case of a cylindrical conductor the field components repsonsible for the inward flow of energy are E_z, the longitudinal component of the electric field, and H_φ, the azimuthal component of the magnetic field. In the a.c. case E_z and H_φ may be regarded as components of a plane wave, and s may be regarded as the depth of penetration into the wire.

If one solves the problem properly in circular coordinates (one gets rather unpleasant looking functions; for details see e.g. Ramo, Whinnery, and van Duzer (1965))† it turns out that the plane-wave approximation is good as long as the radius is large in comparison with the skin depth.

In a practical case there are usually (with the exception of surface-wave lines) two wires, and if the two wires are near to each other the magnetic field on the surfaces will be considerably distorted. Nevertheless, eqn (5.107(a)) provides a very good rule of thumb for working out the current-carrying cross-section of a cylindrical wire.

The phenomenon that currents are confined to the outer skin of the conductor is known as the skin effect. It is, incidentally, harmful. Part of the conductor is there (representing a burden both on the supporting poles and on the balance of payments) without doing anything good.

5.12. Radiation pressure

An electromagnetic wave incident upon a conducting medium produces a force on it. Why? The reason is quite simple. A mobile charge in the

† S. RAMO, J. R. WHINNERY, and T. VAN DUZER (1965). *Fields and waves in communication electronics* p. 291 Wiley New York.

conductor acquires a velocity **v** due to the presence of the electric field **E**. The magnetic field being perpendicular to **E** is perpendicular to **v** as well and will therefore produce a $q\mathbf{v}\times\mathbf{B}$ force acting in the direction of propagation. Since both **v** and **B** vary harmonically as a function of time there will be an average force (same kind as the time average of the Poynting vector) obtainable by working out $\frac{1}{2}\,\mathrm{Re}(\mathbf{v}\times\mathbf{B}^*)$. This force per unit surface, when summed over all the charges present, is known as radiation pressure. The calculation is not too laborious for simple geometries. If you are interested attempt Example 5.6.

Can radiation pressure be harnessed for engineering purposes? Can we build lightmills in analogy with windmills and watermills? Unfortunately the force turns out to be very small. It is, though, measurable; one may, in fact, determine the electromagnetic power by measuring the radiation pressure on suspended vanes (see A. L. Cullen, Absolute power measurement at microwave frequencies, *Proc. I. E. E.* Part III, 1952).

5.13. Waveguides

In this section we shall investigate what happens when the electromagnetic field is imprisoned with the aid of infinitely conducting materials. In practice, of course, we use metals (mostly copper) which have finite conductivities. In the next section, concerned with resonators, losses will indeed be considered, but for the purpose of finding the variation of electric and magnetic fields in a waveguide we shall assume infinitely conducting boundaries.

The simplest waveguide has a rectangular cross-section as shown in Fig. 5.7. The electromagnetic energy propagates down the waveguide along the z direction.

How could we determine the electromagnetic fields in a waveguide? We need to solve Maxwell's equations subject to the boundary conditions. This sounds like a lot of mathematical work. Well, it will take some time to derive the fields, but I believe it will be worth the effort. It is a derivation that brings

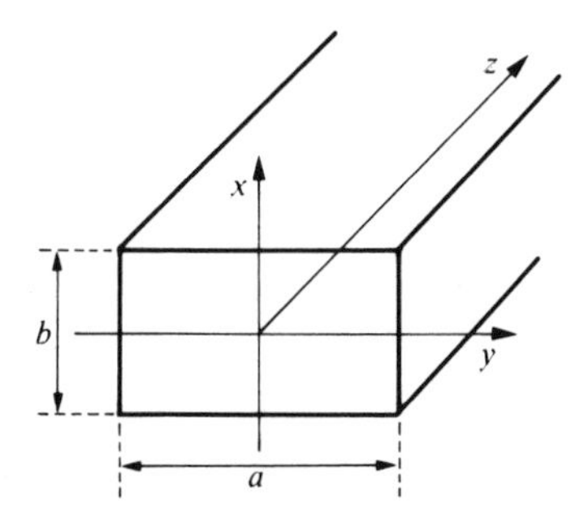

FIG. 5.7. A rectangular waveguide.

home the power of Maxwell's equations and gives support to my often repeated claim that physical intuition is built on mathematical foundations. In this case, of course, we are starting with *some* physical intuition based on empirical evidence. Take a piece of straight rectangular (other cross-sections would do as well, but let's be concrete) waveguide, hold one end to your eye and point the other end towards the sun. If you can see the sun that's a proof that electromagnetic waves can travel in a metal pipe.

It would be possible to look for a general solution of Maxwell's equations in a rectangular waveguide. I feel that general solutions are extremely valuable to have, once you have acquired a certain familiarity with the subject under investigation, but for the first attempt I would rather restrict generality and look for a simple solution on the following lines. By analogy with our plane-wave solution we could first ask the question whether a solution exists in which the only non-zero components are E_x and H_y and there is no variation in the transverse plane. There appears to be no difficulty with the $\partial/\partial x = 0$ condition. E_x can be independent of x, and the boundary conditions on the $x = \pm b/2$ walls will be automatically saisfied. Can E_x be independent of y? Certainly not; E_x must vary as a function of y so as to vanish at $y = \pm a/2$ (remember, the tangential component of the electric field must vanish on the infinitely conducting boundary). But if E_x is a function of y then according to eqn (5.72) there must be an H_y component as well. So the provisional conclusion is that a solution exists in terms of E_x, H_y, and H_z.

Are we entitled to make these assumptions? We are entitled to make any assumptions we like. If they represent a possible solution we will get it. If we made a silly mistake by assuming something self-contradictory, we shall not be rewarded with a solution. So let us see which is the case. Assuming that the waveguide is filled with vacuum, and no currents flow inside, our basic equations ((5.71) and (5.72)) modify to

$$\nabla \times \mathbf{H} = \mathrm{j}\omega\varepsilon_0 \mathbf{E} \tag{5.108}$$

and

$$\nabla \times \mathbf{E} = -\mathrm{j}\omega\mu_0 \mathbf{H}. \tag{5.109}$$

In view of our assumptions the left-hand side of the above equations take the form

$$\nabla \times \mathbf{H} = \begin{vmatrix} \mathbf{i}_x & \mathbf{i}_y & \mathbf{i}_z \\ 0 & \dfrac{\partial}{\partial y} & \dfrac{\partial}{\partial z} \\ 0 & H_y & H_z \end{vmatrix} = \mathbf{i}_x\left(\frac{\partial H_z}{\partial y} - \frac{\partial H_y}{\partial z}\right) \tag{5.110}$$

and

$$\nabla\times\mathbf{E}=\begin{vmatrix}\mathbf{i}_x & \mathbf{i}_y & \mathbf{i}_z\\ 0 & \dfrac{\partial}{\partial y} & \dfrac{\partial}{\partial z}\\ E_x & 0 & 0\end{vmatrix}=\frac{\partial E_x}{\partial z}\mathbf{i}_y-\frac{\partial E_x}{\partial y}\mathbf{i}_z, \tag{5.111}$$

leading to the scalar differential equations

$$\frac{\partial E_x}{\partial z}=-\mathrm{j}\omega\mu_0 H_y, \tag{5.112}$$

$$\frac{\partial E_x}{\partial y}=\mathrm{j}\omega\mu_0 H_z, \tag{5.113}$$

$$\frac{\partial H_z}{\partial y}-\frac{\partial H_y}{\partial z}=\mathrm{j}\omega\varepsilon_0 E_x. \tag{5.114}$$

Expressing H_y and H_z from eqns (5.112) and (5.113), and substituting them into eqn (5.114), we get

$$\frac{\partial^2 E_x}{\partial y^2}+\frac{\partial^2 E_x}{\partial z^2}+\beta_0^2 E_x=0. \tag{5.115}$$

Attempting the solution in the form

$$E_x=Y(y)Z(z), \tag{5.116}$$

we get

$$Z\frac{\mathrm{d}^2 Y}{\mathrm{d}y^2}+Y\frac{\mathrm{d}^2 Z}{\mathrm{d}z^2}+\beta_0^2 YZ=0, \tag{5.117}$$

or, dividing by YZ, we get the modified form

$$\frac{1}{Y}\frac{\mathrm{d}^2 Y}{\mathrm{d}y^2}+\frac{1}{Z}\frac{\mathrm{d}^2 Z}{\mathrm{d}z^2}=-\beta_0^2, \tag{5.118}$$

which can only be satisfied for all values of y and z if both Y''/Y and Z''/Z are separately equal to a constant, namely,

$$\frac{1}{Y}\frac{\mathrm{d}^2 Y}{\mathrm{d}y^2}=-\eta^2 \quad\text{and}\quad \frac{1}{Z}\frac{\mathrm{d}^2 Z}{\mathrm{d}z^2}=-\beta^2, \qquad \beta^2+\eta^2=\beta_0^2. \tag{5.119}$$

I hope you find these differential equations familiar; the solutions are trigonometric and exponential functions. Since the wave may be expected to propagate in the z direction, the solution for Z will be written as

$$Z=Z_{\mathrm{f}}\,\mathrm{e}^{-\mathrm{j}\beta z}+Z_{\mathrm{r}}\,\mathrm{e}^{\mathrm{j}\beta z}, \tag{5.120}$$

where Z_f and Z_r are constants. No propagation is expected in the y direction, hence we shall choose a trigonometric form, as follows

$$Y = Y_1 \cos \eta y + Y_2 \sin \eta y, \tag{5.121}$$

where Y_1 and Y_2 are constants. The general solution is then

$$E_x = (Y_1 \cos \eta y + Y_2 \sin \eta y)(Z_f\, e^{-j\beta z} + Z_r\, e^{j\beta z}). \tag{5.122}$$

The boundary conditions

$$E_x = 0 \quad \text{at } y = \pm a/2 \tag{5.123}$$

are satisfied with the following choice of the constants

$$Y_2 = 0, \qquad Y_1 Z_f = E_f, \qquad Y_1 Z_r = E_r, \qquad \eta = \pi/a \tag{5.124}$$

leading to

$$E_x = \cos \frac{\pi y}{a}(E_f\, e^{-j\beta z} + E_r\, e^{j\beta z}). \tag{5.125}$$

Let's see what we have got. We have two waves travelling in the direction of the positive and negative z axis respectively, the same kind of solution we obtained before for plane waves. There is though one essential difference. The propagation coefficient is given now as

$$\beta = (\beta_0^2 - \eta^2)^{\frac{1}{2}} = \left(\frac{\omega^2}{c^2} - \frac{\pi^2}{a^2}\right)^{\frac{1}{2}}, \tag{5.126}$$

dependent on the dimension a of the waveguide. If the frequency is low enough then β becomes imaginary and no propagation is possible. The frequency at which this happens is called the cut-off frequency, obtained from the equation

$$\frac{\omega_c^2}{c^2} - \frac{\pi^2}{a^2} = 0, \tag{5.127}$$

yielding

$$f_c = \frac{c}{2a} \quad \text{or} \quad \lambda_c = 2a. \tag{5.128}$$

This is a formula easy to remember. As long as the half-wavelength in free space is smaller than the dimension of the guide in the y direction, propagation is possible.

What is the wavelength in the waveguide (or rather along the waveguide)? By the definition of wavelength as the period in space, we get

$$\lambda_g = \frac{2\pi}{\beta} = \frac{2\pi}{\{(2\pi/\lambda)^2 - (\pi/a)^2\}^{\frac{1}{2}}} = \frac{\lambda}{\{1 - (\lambda/\lambda_c)^2\}^{\frac{1}{2}}}. \tag{5.129}$$

What happens when $\lambda > \lambda_c$? As I have just said, there will be no propagation. The mathematical answer follows clearly from eqn (5.125). When β becomes imaginary the amplitude of the field decreases exponentially.

The magnetic field may be obtained from eqns (5.112) and (5.113) as follows:

$$H_y = -\frac{1}{j\omega\mu_0}\frac{\partial E_x}{\partial z} = \frac{\beta}{\omega\mu_0}\cos\frac{\pi y}{a}(E_f\,e^{-j\beta z} - E_r\,e^{j\beta z}); \tag{5.130}$$

$$H_z = \frac{1}{j\omega\mu_0}\frac{\partial E_x}{\partial y} = j\frac{\pi}{\omega\mu_0 a}\sin\frac{\pi y}{a}(E_f\,e^{-j\beta z} + E_r\,e^{j\beta z}). \tag{5.131}$$

We have now got both the electric and magnetic fields. At a given moment in time, in the transverse plane the electric and magnetic field lines are shown in Fig. 5.8.

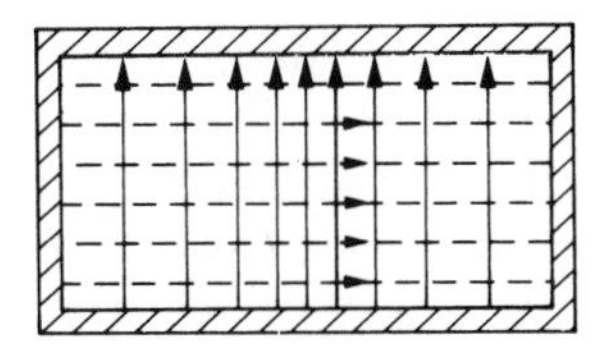

FIG. 5.8. Electric and magnetic field lines.

As you may have guessed, it was no coincidence that I suggested an E_x component for the electric field, and no variation in the x direction ($\partial/\partial x = 0$). These assumptions led not only to the *simplest* field distribution but also to the one used in practice. As you may have noticed the choice of the constants in eqn (5.124) was a rather specific one leading to the desired result. There were other possibilities too (e.g. $Y_1 = 0$, $\eta = 2m\pi/a$, $m = 1, 2 \ldots$) all exhibiting solutions mathematically possible and physically realizable.

Each solution that satisfies the boundary conditions is called a mode. The one we have derived is the lowest mode. In other words, it is the first allowed mode. At large wavelengths (low frequencies) there is no propagation. As we lower the wavelength (increase the frequency) there is a certain point at which λ becomes less than $\lambda_c = 2a$, and our mode can cheerfully propagate down the waveguide (E_x as a function of y for this mode is shown in Fig. 5.9(a)). No other modes are possible. However, when the wavelength is further decreased there is now room for a whole sinusoidal period (Fig. 5.9(b)) which we could call mode 2, the second allowed mode. The larger is the waveguide relative to the wavelength the more modes are permitted, and that applies of course not only to the y direction but to the x direction as well.

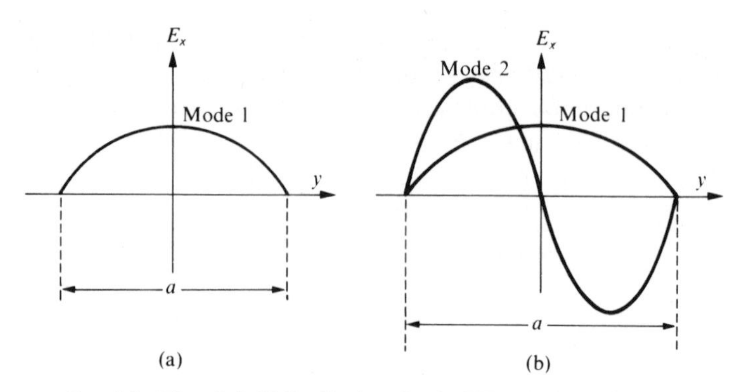

FIG. 5.9. Electric field distributions for the TE_{10} and TE_{20} modes.

Is it possible to show the existence of these higher modes mathematically? Yes, certainly. We may, for example, obtain a set of new solutions by permitting an E_y component (in addition to E_x) and by dropping our insistence on the condition $\partial/\partial x = 0$. The solutions obtained that way are called TE modes because the electric field is always in the transverse plane. Since there are so many of these modes, an indexing system is introduced. In a TE_{mn} (see Example 5.8) mode the electric field is described by m half-sinusoidals in the y direction and by n half-sinusoidals in the x direction. Hence the field variations shown in Fig. 5.9(b) would correspond to that of the TE_{10} and TE_{20} modes.

There are of course *TM* (*transverse magnetic*) *modes* as well where all the components of the magnetic field are in the transverse plane. I am afraid I have no time to discuss here the general theory of waveguides; for an excellent review see Simonyi (1963).† My main aim was to show that at high enough frequencies electricity is capable to flow inside a closed metal structure. If you want an idea of the actual size of a practical waveguide, note that most radar works at 10 GHz, leading to a waveguide size of above the half-wavelength but below the full wavelength, say 25 mm. If you want to get the 10 GHz power from the tube that produces it to the aerial that radiates it out, this is the waveguide you will use.

Are all waveguides rectangular? Well, in principle, waveguide cross-section could be of any shape whatsoever (some mathematically minded engineers have great fun working out the field distribution for all sorts of odd-looking cross-sections). The next logical choice for the cross-section is a circle. Circular waveguides are indeed used in practice, but only in special circumstances. The most important special circumstance is the existence of a low-loss mode‡ not found for any other cross-section.

† K. SIMONYI (1963). *Foundations of electrical engineering*, pp. 713–58. Pergamon Press, Oxford.

‡ The British Post Office used to have an experimental line 23 miles long using this low-loss mode. They hoped to use it for inter-city communications. Alas, the microwave waveguide was unable to compete with optical fibres which came into the race later. The experiments were abandoned.

We shall now derive the mathematical solution for this low-loss mode in a manner similar to that used for the rectangular waveguide but employing cylindrical coordinates.

We shall assume circular symmetry ($\partial/\partial\varphi = 0$) and the existence of three non-zero components E_φ, H_r, and H_z. Starting again with eqns (5.108) and (5.109) we need to evaluate the curls

$$\nabla\times\mathbf{H} = \frac{1}{R}\begin{vmatrix} \mathbf{i}_R & R\mathbf{i}_\varphi & \mathbf{i}_z \\ \dfrac{\partial}{\partial R} & 0 & \dfrac{\partial}{\partial z} \\ H_R & 0 & H_z \end{vmatrix} = -\mathbf{i}_\varphi\left(\frac{\partial H_z}{\partial R} - \frac{\partial H_R}{\partial z}\right) \tag{5.132}$$

and

$$\nabla\times\mathbf{E} = \frac{1}{R}\begin{vmatrix} \mathbf{i}_R & R\mathbf{i}_\varphi & \mathbf{i}_z \\ \dfrac{\partial}{\partial R} & 0 & \dfrac{\partial}{\partial z} \\ 0 & RE_R & 0 \end{vmatrix} = -\mathbf{i}_R\frac{\partial E_\varphi}{\partial z} + \mathbf{i}_z\frac{1}{R}\frac{\partial}{\partial R}(RE_\varphi), \tag{5.133}$$

leading to the scalar differential equations

$$\frac{\partial E_\varphi}{\partial z} = \mathrm{j}\omega\mu H_R, \tag{5.134}$$

$$\frac{1}{R}\frac{\partial}{\partial R}(RE_\varphi) = -\mathrm{j}\omega\mu H_z, \tag{5.135}$$

$$\frac{\partial H_z}{\partial R} - \frac{\partial H_R}{\partial z} = -\mathrm{j}\omega\varepsilon E_\varphi, \tag{5.136}$$

which yield, after a bit of work,

$$\frac{\partial^2 E_\varphi}{\partial R^2} + \frac{1}{R}\frac{\partial E_\varphi}{\partial R} + \left(\beta_0^2 - \frac{1}{R^2}\right)E_\varphi = 0. \tag{5.137}$$

Attempting the solution in the form

$$E_\varphi = G(R)Z(z) \tag{5.138}$$

eqn (5.137) may be transformed into

$$\frac{1}{G}\frac{\mathrm{d}^2 G}{\mathrm{d}R^2} + \frac{1}{R}\frac{1}{G}\frac{\mathrm{d}G}{\mathrm{d}R} - \frac{1}{R^2} + \frac{1}{Z}\frac{\mathrm{d}^2 Z}{\mathrm{d}z^2} = -\beta_0^2. \tag{5.139}$$

Equating the z-dependent term to $-\beta^2$ we get

$$\frac{\mathrm{d}^2 Z}{\mathrm{d}z^2} + \beta^2 Z = 0, \tag{5.140}$$

giving the solution

$$Z = Z_f\,e^{-j\beta z} + Z_r\,e^{j\beta z}. \tag{5.141}$$

There is nothing new so far. We have seen this type of solution before.

The differential equation in G looks more awkward. If we equate the R-dependent terms in eqn (5.128) to $-\zeta^2$ so that

$$\zeta^2 + \beta^2 = \beta_0^2, \tag{5.142}$$

we get the differential equation

$$\frac{d^2G}{dR^2} + \frac{1}{R}\frac{dG}{dR} + \left(\zeta^2 - \frac{1}{R^2}\right)G = 0. \tag{5.143}$$

This is a differential equation, I believe, that you have not come across, as yet. What should one do when coming across a new differential equation? Try to solve it by all means, but if the first ten minutes do not lead to success, it is best to consult a reference book. Since it would take some time to find your way around, I'd rather give a little help. Take the classical book of Jahnke and Emde (revised recently by Lösch),† and on page 156 you will find the differential equation (I am changing the original notations so as not to clash with ours)

$$\frac{d^2w}{dr^2} + \frac{1}{r}\frac{dw}{dr} + \left(1 + \frac{\nu^2}{r^2}\right)w = 0. \tag{5.144}$$

We are nearly there. Let us introduce the new independent variable

$$r = \zeta R, \tag{5.145}$$

then

$$\frac{d}{dR} = \zeta\frac{d}{dr} \quad \text{and} \quad \frac{d^2}{dR^2} = \zeta^2\frac{d^2}{dr^2}, \tag{5.146}$$

leading to

$$\frac{d^2G}{dr^2} + \frac{1}{r}\frac{dG}{dr} + G\left(1 - \frac{1}{r^2}\right) = 0, \tag{5.147}$$

and this is a special case of eqn (5.144) with $\nu = 1$. The solution is the so-called Bessel function of first order, $J_1(r)$.

There is nothing to be afraid of; Bessel functions are not more difficult than other functions. $J_1(r)$ is defined in terms of an infinite power series, there are approximate formulae for small r and for large r, it is available (at

† JAHNKE–EMDE–LÖSCH, *Tables of higher functions*, Teubner, Stuttgart, 1960. It provides an excellent opportunity for extending your knowledge of foreign languages. One side of each page is written in German, the other side in English.

least as a subroutine) in better computers, and it is tabulated in the above-mentioned book (and in many others).

At $R = R_0$ (where R_0 is the radius of the circular waveguide) E_φ must vanish, leading to the mathematical condition that ζR_0 must be equal to one of the roots of $J_1(r)$. In the simplest case we can take the first root at $r_1 = 3{\cdot}832$ leading to

$$\zeta = \frac{3{\cdot}832}{R_0} \quad \text{and} \quad \beta = \left\{\beta_0^2 - \left(\frac{3{\cdot}832}{R_0}\right)\right\}^{\frac{1}{2}}. \tag{5.148}$$

With ζ and β defined as above, the electric field in this particular mode (called TE_{01} mode because it has no zeros in the azimuthal direction and one in the radial direction) may be expressed as

$$E_\varphi = J_1(r)(E_f\,e^{-j\beta z} + E_r\,e^{j\beta z}), \tag{5.149}$$

and a little more mathematical effort would yield the components of the magnetic field as well (see Example 5.14).

The field lines for the electric† and magnetic fields are illustrated in Fig. 5.10(a); E_φ and H_R vary in the same manner as a function of normalized radius as plotted in Fig. 5.10(b), whereas H_z is shown in Fig. 5.10(c).

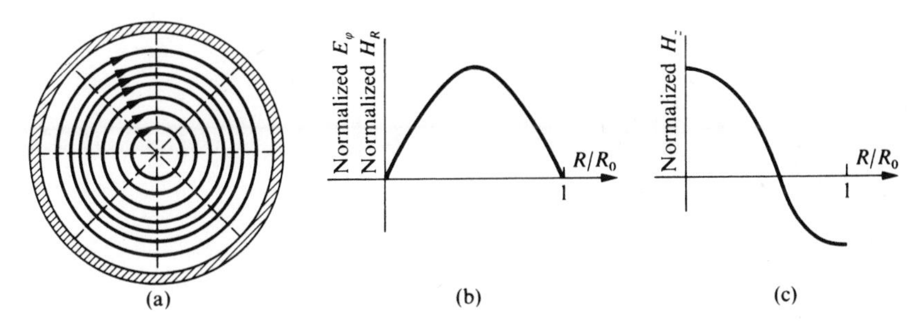

FIG. 5.10. Field distributions of the TE_{01} mode in a circular waveguide.

Of course, everything I said about the rectangular waveguide applies to the circular waveguide as well. There are TE and TM modes, and each mode has a cut-off wavelength. Note that the TE_{01} mode in the circular waveguide is *not* the lowest mode. A number of others can coexist, giving frequent headaches to those engineers who want to use the waveguide for long-distance communications.

Next, we shall work out the field distribution for the oldest waveguide in use, the coaxial waveguide (perhaps better known as a coaxial cable). It

† There is now some indication why the TE_{01} mode is a low loss mode. The electric field has no normal components at the waveguide wall. If you are not convinced by this simple argument I am afraid you have to solve Examples 5.11–5.15.

consists of an inner and an outer conductor as shown in Fig. 5.11. What is the simplest mode one can think of? Intuition based on electrostatics and magnetostatics suggests that there should be a circularly symmetric solution with a radial electric field and an azimuthal magnetic field (Fig. 5.11(a)). Any axial components? At first sight they do not appear to be necessary. The test

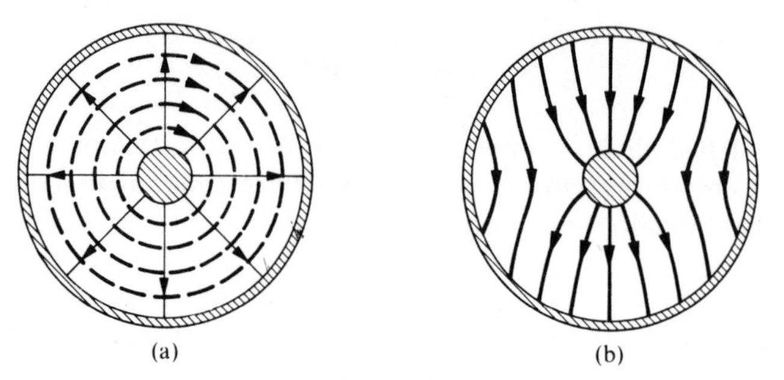

FIG. 5.11. (a) The usual TEM and (b) the TE_{11} mode in a coaxial waveguide.

is, of course, whether they can satisfy the differential equations (the boundary conditions are already satisfied). So let us take the curls again in a cylindrical coordinate system.

$$\nabla \times \mathbf{H} = \frac{1}{R}\begin{vmatrix} \mathbf{i}_R & R\mathbf{i}_\varphi & \mathbf{i}_z \\ \frac{\partial}{\partial R} & 0 & \frac{\partial}{\partial z} \\ 0 & RH_\varphi & 0 \end{vmatrix} = -\mathbf{i}_R \frac{\partial H_\varphi}{\partial z} + \mathbf{i}_z \frac{1}{R}\frac{\partial}{\partial R}(RH_\varphi) \tag{5.150}$$

and

$$\nabla \times \mathbf{E} = \frac{1}{R}\begin{vmatrix} \mathbf{i}_R & R\mathbf{i}_\varphi & \mathbf{i}_z \\ \frac{\partial}{\partial R} & 0 & \frac{\partial}{\partial z} \\ E_R & 0 & 0 \end{vmatrix} = \mathbf{i}_\varphi \frac{\partial E_R}{\partial z}, \tag{5.151}$$

leading to the scalar differential equations

$$-\frac{\partial H_\varphi}{\partial z} = \mathrm{j}\omega\varepsilon E_R, \tag{5.152}$$

$$\frac{\partial}{\partial R}(RH_\varphi) = 0, \tag{5.153}$$

$$\frac{\partial E_R}{\partial z} = -\mathrm{j}\omega\mu H_\varphi. \tag{5.154}$$

It may be easily derived from the above differential equations that both E_R and H_φ vary as $1/R$ and that the z variation may be obtained from the differential equation

$$\frac{\partial^2 E_R}{\partial z^2} + \beta_0^2 E_R = 0. \tag{5.155}$$

Note that the above equation has a β_0 not dependent on the dimension of the waveguide, hence the guide wavelength is equal to the free-space wavelength. The propagation is similar to that of a plane wave, offering another example of a TEM wave.

Is there a cut-off wavelength in a coaxial waveguide? Obviously not. You can see that the propagation coefficient is real for all frequencies or if you still prefer the old-fashioned low-frequency concepts then you can argue that there are two conductors, one to carry the current there and one to bring it back. Hence the coaxial waveguide should be all right for d.c. as well. Are there any higher modes in a coaxial waveguide? Yes, an infinite number of them. The first higher-order mode appears when the wavelength is comparable with the circumference of the outer conductor. I shall show you the field distribution of this mode in Fig. 5.11(b) just to dispel the notion (held by many) that the TEM mode is the only possible one in a coaxial waveguide. Are there any other simple structures capable of carrying a TEM mode? I haven't yet mentioned the simplest one, which consists of two infinitely large sheets (located at x_1 and x_2 in Fig. 5.12). Our plane wave described by eqn

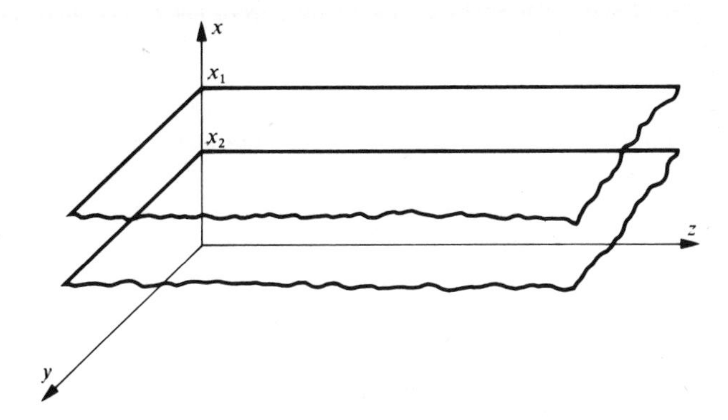

FIG. 5.12. TEM waves propagate between two infinitely large metal plates.

(5.77) provides the solution. It satisfies the boundary conditions (electric field is perpendicular to the plates), so it can propagate between the plates. It is as if we took a slice of an infinitely large plane wave.

The main disadvantage of two infinitely large parallel plates is their infinite extent. Could we reduce their transverse dimensions and obtain a waveguide as shown in Fig. 5.13(a)? This is indeed possible. The family of guiding structures based on this idea comes under the heading of strip lines. This is the time to apply your freshly acquired physical intuition. I am not going to solve any equations but I hope you will agree that the field lines must look something like those shown in Fig. 5.13(b). The obvious trouble is that the fields are no longer imprisoned; they can spread out. How could we confine them? As long as we have an open structure the field lines will always leak out, but it is possible to improve the situation by filling the space between the plates by a material of high relative permittivity (incidentally that also helps to hold the upper conductor in place). If we want to put several strip lines not too far from each other then it is preferable to have a conducting ground plane with a dielectric on top, and then only the upper conductor needs to be put to the right place. A modern strip line is shown in Fig. 5.13(c).

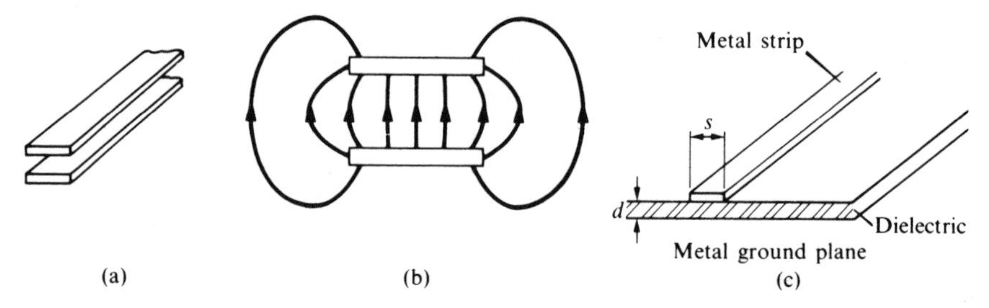

FIG. 5.13. (a) A strip line. (b) Electric field lines. (c) Strip line with ground plane.

What have we gained by using a strip line? Size and economy are the most important considerations. Remember that the size of the rectangular waveguide for use at 10 GHz was 25 mm. A strip line used for the same frequency band would have (Fig. 5.13(c)) $d = s \simeq 0{\cdot}63$ mm, representing a significant reduction in size.† Such construction turns out to be much cheaper as well (even if it is made of gold), because modern mass production techniques (evaporation of the metal, application of photoresists, etching, etc.) developed for low-frequency integrated circuits may be used. Can we get a mathematical solution? Not easily. Since energy can radiate out this is no longer a neat boundary-value problem. A TEM mode does not exist; one may though expect a field distribution not very different from that of a TEM mode.‡ For a given geometry approximate numerical solutions are always possible. One such solution for the electric field is shown in Fig. 5.14. As you can see the behaviour in the z direction is slightly unusual. An electric field

† For being so small this version is usually referred to as a microstrip.

‡ It is easy to show (Example 5.7) that for a TEM mode the electric field distribution in the transverse plane agrees with the static solution (it is Laplace's equation that needs to be solved).

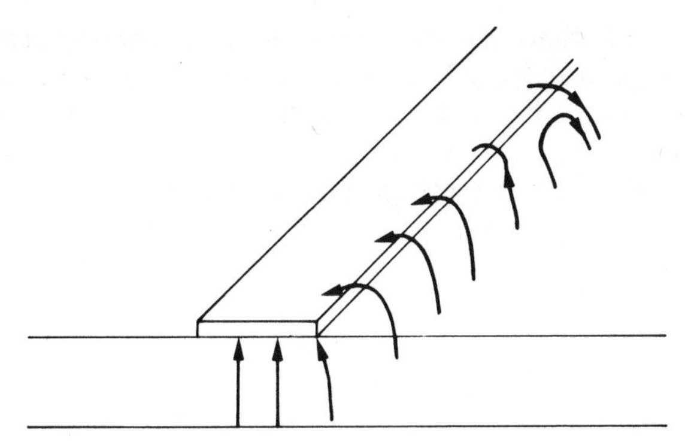

FIG. 5.14. A numerical solution for electric field lines.

line will not necessarily connect the two conductors; it might originate and terminate on the same conductor.

This is about as much as I wanted to say about waveguides. The aim of this section was not to make you experts on waveguides, nor was it my intention to give complete solutions. I have simply tried to convey the message that metal structures (sometimes aided and abetted by dielectrics) are suitable for guiding electromagnetic waves.

5.14. Resonators

If you feel happy about the propagation of electromagnetic waves in free space or in a waveguide then you should have no difficulties with resonators either. As I mentioned before, a high-frequency electromagnetic resonator is much nearer to 'common sense' than its low-frequency counterpart, the *LC* circuit.

Think of standing-wave patterns on a string or in an organ pipe and it becomes obvious that similar phenomena must occur in electromagnetic resonance. We have already come across electromagnetic standing waves when studying the reflection of a plane wave by an infinitely conducting plate (see Fig. 5.4). The standing-wave pattern will obviously undergo no change if we insert an infinitely conducting plate parallel with the first plate at the position of a node where the electric field is zero. We obtain then a resonator in which the electromagnetic energy bounces between the plates until the end of time—provided that the plates are indeed infinitely conducting.

The main trouble with this resonator is that it is too large for practical applications; infinite dimensions are usually frowned upon in engineering designs. Can we get away with finite dimensions? Yes, that's a possibility to which we shall return later in this section. But let's play it safe and choose first a resonator from which the energy cannot leak out. It will take the form of a rectangular waveguide terminated by two plates made of the same material (Fig. 5.15). Since this is an empty box, it is called a *cavity resonator.* It will resonate at a frequency at which l, the length of the cavity, is an integral multiple of the half-wavelength—measured of course in the waveguide.

Take as an example a rectangular waveguide for which $a = 24$ mm, $b = 12$ mm. At a frequency of 10 GHz the guide wavelength in the TE_{10} mode is

$$\lambda_g = \frac{\lambda}{\{1-(\lambda/2a)^2\}^{\frac{1}{2}}} = \frac{30}{\{1-(30/48)^2\}^{\frac{1}{2}}} = 38{\cdot}4 \text{ mm}. \qquad (5.156)$$

Hence the simplest resonator will have a length of 19·2 mm. The resulting cavity mode is, quite logically, referred to as the TE_{101} mode, where the last

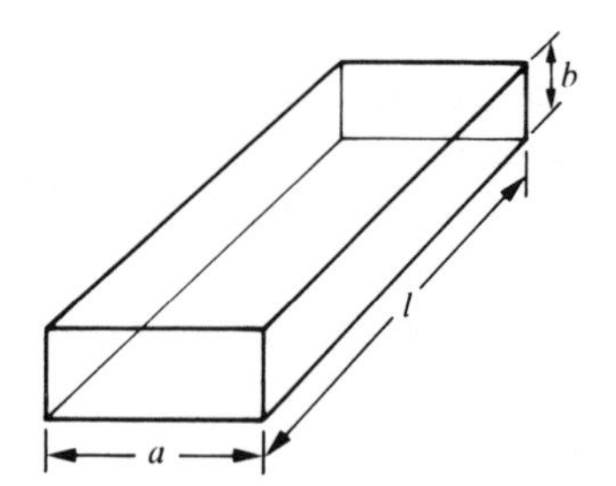

FIG. 5.15. A cavity resonator.

subscript stands in general for the number of half-wavelengths along the length of the cavity.

Naturally, one can make resonators out of coaxial and circular waveguides as well. The principles are the same, though not the guide wavelengths.

Do strip-line resonators exist? A strip of the right length does indeed behave as a resonator, but as a poor resonator because the reflection at the end is far from being perfect. At each transit some of the energy is radiated out. Is it possible to reduce the radiation losses? Yes, one can do that by employing a different design. Instead of running to and fro the electromagnetic energy can travel round and round. This is the idea behind the circular strip-line resonator shown in Fig. 5.16, where the total length (along the dotted line) is an integral multiple of the half-wavelength.

Quite different from strip lines but employing also an open structure are the so-called *open resonators,* which consist of two reflectors of finite size facing each other (Fig. 5.17 shows two reflectors having flat and spherical shapes respectively). How large should the metal plates be? Whenever wave phenomena are concerned, the general rule to remember is that large or small always refer to comparisons with the wavelength. A large plate would mean many, many wavelengths. How many? Say a thousand or at least a

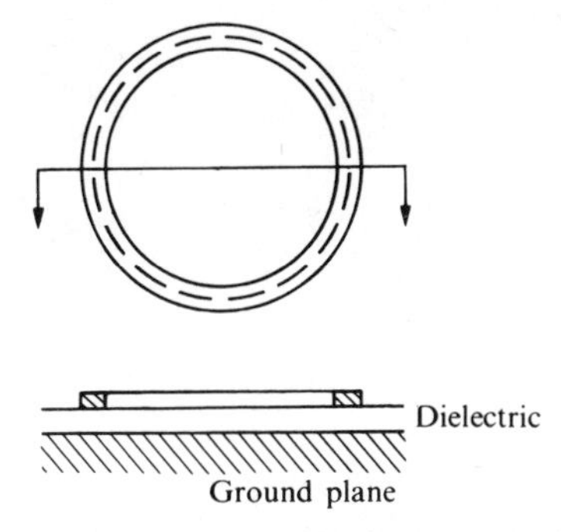

FIG. 5.16. A circular strip-line resonator.

hundred. Take again a radar set working at 10 GHz, then a plate of hundred wavelengths diameter would measure 3 m, obviously not a practical proposition. But if we move towards shorter wavelengths, say down to 0·5 μm, then the situation is markedly better. A diameter of 5 mm is 10 000 times larger than the wavelength. Hence open resonators should come in useful at optical wavelengths, and indeed that is the standard form of resonators in most lasers.

Is there a mode structure in open resonators? Yes, one can make nice approximations and find a set of TEM modes. The simplest solution may be obtained for the spherical mirrors of Fig. 5.17(b). It is possible to show (the mathematical labour involved is not too strenuous†) that Maxwell's equations permit a solution in which a wave can propagate back and forth

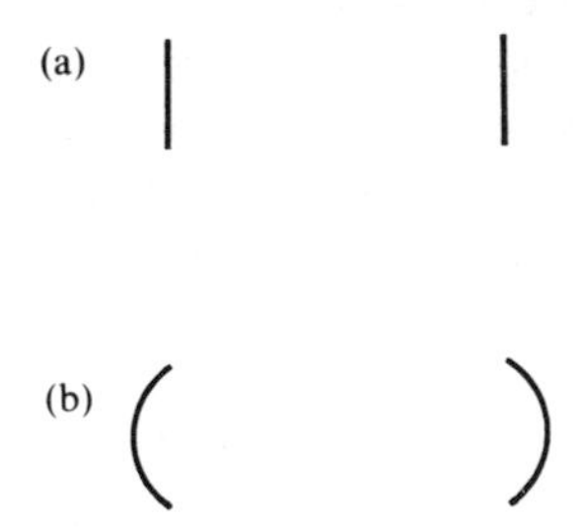

FIG. 5.17. Open resonators. (a) Plane, (b) spherical reflectors.

between the reflectors without any change of shape. Such a stable solution has got the right to be called a mode. The lowest mode (TEM_{00}) has cylindrical symmetry; its amplitude is maximum at the axis and decays in the radial direction as a gaussian function. The solid lines in Fig. 5.18 represent

† H. KOGELNIK and T. LI (1966). Laser beams and resonators. *Proc. IEEE* **54** 1312–29.

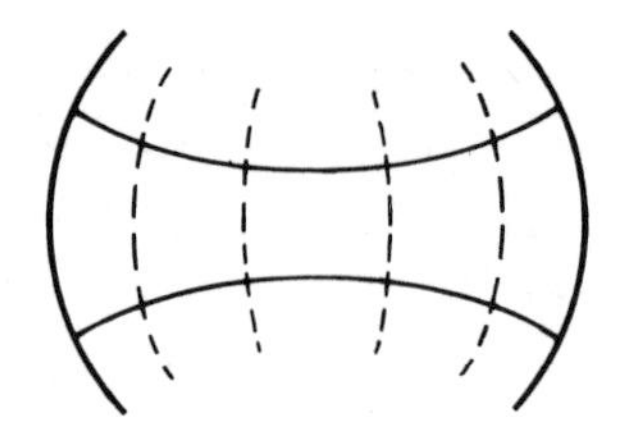

FIG. 5.18. Rays and phasefronts in an open resonator with spherical reflectors.

constant amplitudes whereas the dotted lines give the phasefronts. Note that the reflector itself must coincide with a constant-phase surface. No such 'nice' solution exists for the plane reflectors of Fig. 5.17(a), but it is still true that under resonance conditions the field distribution remains unchanged.

I have now mentioned most electromagnetic resonators of practical importance. In conclusion I would like to say a few words on the *quality* of resonators. The definition is the same as for the quality factor of any kind of resonator, namely

$$Q = 2\pi \frac{\text{energy stored}}{\text{energy lost per cycle}}, \tag{5.157}$$

where the energy may be lost either by dissipation or by radiation (for a formula in a concrete case see Example 5.21).

It is pretty difficult to work out radiation losses (that must be left to a more specialized course), but I can outline the calculation of resistive losses. First of all one has to admit (as in Section 5.11) that the walls are made of real materials, usually copper, silver, or gold, which do not have infinitely large conductivities. However, the losses are small, so that our previous determination of field distributions may still be regarded valid. In fact the only change we need to make is that the surface currents (remember $H_t = K$ from the boundary conditions, eqns (5.54)) flow now in a resistive medium, giving rise to losses. For an alternative explanation in terms of power moving into the walls, see Examples 5.13–15.

5.15. Reflection and refraction at the interface of two dielectric media

We shall in this section return to the investigation of plane waves, namely to their reflection and refraction at a plane interface between two dielectric media. Assume that the space left to the $z = 0$ plane is filled with dielectric 1 and that to the right by dielectric 2 having relative permittivities ε_1 and ε_2 respectively. We shall take the conductivities to be zero and the permeabilities to be equal to that in vacuum. It wouldn't be much effort to be more general and take σ_1 and σ_2 for conductivities and μ_1 and μ_2 for relative permeabilities, but I always prefer to attack a simpler problem first so let us stick to lossless dielectrics.

We shall first look at the case when the magnetic field vector is parallel with the interface, the same choice for field vectors as in Section 5.10, where the reflection by a perfectly conducting surface was investigated. Hence the electric field in the incident wave takes the form

$$\mathbf{E}_i = E_0(\cos\alpha_1\mathbf{i}_x - \sin\alpha_1\mathbf{i}_z)\exp\{-j\beta_1(x\sin\alpha_1 + z\cos\alpha_1)\}, \quad (5.158)$$

where $\beta_1 = \omega\sqrt{(\mu_0\varepsilon_1)}$ is the propagation coefficient in medium 1. We may still assume that the angle of reflection is equal to the angle of incidence giving the reflected electric field in the form

$$\mathbf{E}_r = \Gamma E_0(-\cos\alpha_1\mathbf{i}_x - \sin\alpha_1\mathbf{i}_z)\exp\{-j\beta_1(x\sin\alpha_1 - z\cos\alpha_1)\}. \quad (5.159)$$

Note that only a part of the incident energy will be reflected, so we had to introduce a reflection coefficient $|\Gamma| < 1$. One may similarly introduce a transmission coefficient for the wave travelling in medium 2, leading to the following expression for the electric field

$$\mathbf{E}_T = TE_0(\cos\alpha_2\mathbf{i}_x - \sin\alpha_2\mathbf{i}_z)\exp\{-j\beta_2(x\sin\alpha_2 + z\cos\alpha_2)\}, \quad (5.160)$$

where α_2 is the angle of refraction (Fig. 5.19(a)).

For a given incident angle α_1 there are three unknowns to be determined, α_2, Γ, and T. They may be obtained from the boundary conditions demanding that the tangential components of the electric field and the normal

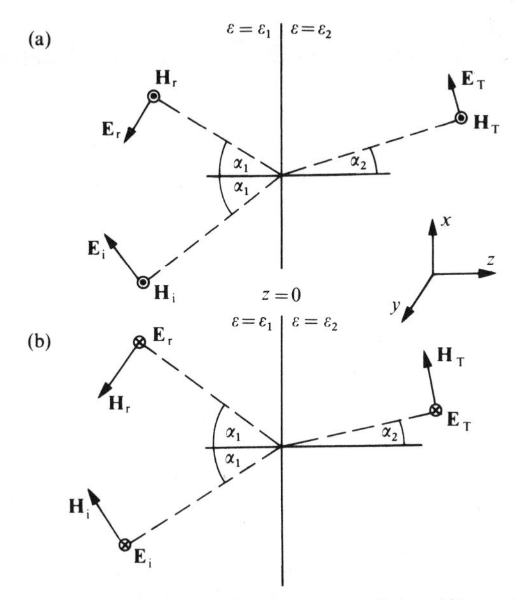

FIG. 5.19. Reflection and refraction of plane waves at a dielectric boundary. (a) Magnetic, (b) electric field parallel with the interface.

components of the electric flux density must be continuous at the interface,† i.e.

$$(\mathbf{E}_i+\mathbf{E}_r)_t=(\mathbf{E}_T)_t \quad \text{and} \quad \varepsilon_1(\mathbf{E}_i+\mathbf{E}_r)_n=\varepsilon_2(\mathbf{E}_T)_n. \tag{5.161}$$

Substituting eqn (5.158) to eqn (5.160) into eqn (5.161) and taking $z=0$ we get

$$(1-\Gamma)E_0\cos\alpha_1\exp(-j\beta_1 x\sin\alpha_1)\equiv TE_0\cos\alpha_2\exp(-j\beta_2 x\sin\alpha_2) \tag{5.162}$$

and

$$-\varepsilon_1 E_0(1+\Gamma)\sin\alpha_1\exp(-j\beta_1 x\sin\alpha_1)=-\varepsilon_2 TE_0\sin\alpha_2\exp(-j\beta_2 x\sin\alpha_2). \tag{5.163}$$

Since the above equations must be satisfied for all values of the x coordinate, it follows that

$$\beta_1\sin\alpha_1=\beta_2\sin\alpha_2 \tag{5.164}$$

or, in other form,

$$\frac{\sin\alpha_1}{\sin\alpha_2}=\sqrt{\left(\frac{\varepsilon_2}{\varepsilon_1}\right)}, \tag{5.165}$$

which you may recognize as Snell's law if we introduce the refraction index n with the relation

$$n=\sqrt{\varepsilon_r}. \tag{5.166}$$

The reflection and transmission coefficients may now be determined from eqns (5.162) and (5.163), yielding

$$\Gamma_\mathbf{H}=\frac{\sqrt{\varepsilon_2}\cos\alpha_1-\sqrt{\varepsilon_1}\cos\alpha_2}{\sqrt{\varepsilon_2}\cos\alpha_1+\sqrt{\varepsilon_1}\cos\alpha_2}=\frac{Z_1\cos\alpha_1-Z_2\cos\alpha_2}{Z_1\cos\alpha_1+Z_2\cos\alpha_2} \tag{5.167}$$

and

$$T_\mathbf{H}=\frac{2\sqrt{\varepsilon_1}\cos\alpha_1}{\sqrt{\varepsilon_2}\cos\alpha_1+\sqrt{\varepsilon_1}\cos\alpha_2}=\frac{2Z_2\cos\alpha_1}{Z_1\cos\alpha_1+Z_2\cos\alpha_2}, \tag{5.168}$$

where $Z_1=(\mu_0/\varepsilon_1)^{\frac{1}{2}}$ and $Z_2=(\mu_0/\varepsilon_2)^{\frac{1}{2}}$, in agreement with the definition of eqn (5.65). The subscript $\mathbf{H}$ was introduced to emphasize that this is a solution for the case when the magnetic field is parallel to the interface.

† It is not entirely obvious which two of the four boundary conditions (eqn (5.52)) to take into account. In the present case it would have been equally satisfactory to match the tangential components of the electric and magnetic fields; on the other hand demanding the continuity of B_n would not have supplied us with an equation for the simple reason that $B_n=0$. I can't give you a general rule how to set about satisfying the boundary conditions. Rely on your physical intuition and choose one of them, and continue afterwards until you have enough equations to determine the unknown coefficients. The rest of the boundary conditions (if there are any left) will then be automatically satisfied. If you come to a contradiction, I mean if you have run out of coefficients and some of the boundary conditions are still not satisfied, then look at your assumptions and approximations again.

The reflection and transmission coefficients may be derived in an entirely analogous manner when the electric field is assumed to be parallel to the interface (Fig. 5.19(b)). It is then more convenient to work in terms of the magnetic field and write the field vectors as follows

$$\mathbf{H}_i = H_0(\cos\alpha_1\mathbf{i}_x - \sin\alpha_1\mathbf{i}_z)\exp\{-j\beta_1(x\sin\alpha_1 + z\cos\alpha_1)\}, \qquad (5.169)$$

$$\mathbf{H}_r = \Gamma_E H_0(-\cos\alpha_1\mathbf{i}_x - \sin\alpha_1\mathbf{i}_z)\exp\{-j\beta_1(x\sin\alpha_1 - z\cos\alpha_1)\}, \qquad (5.170)$$

$$\mathbf{H}_T = T_E H_0(\cos\alpha_2\mathbf{i}_x - \sin\alpha_2\mathbf{i}_y)\exp\{-j\beta_2(x\sin\alpha_2 + z\cos\alpha_2)\}. \qquad (5.171)$$

We may now use the continuity of H_t and B_n on the boundary for determining the unknown coefficients. Snell's law is of course valid in this case as well, whereas for Γ_E and T_E we obtain

$$\Gamma_E = \frac{\sqrt{\varepsilon_1}\cos\alpha_1 - \sqrt{\varepsilon_2}\cos\alpha_2}{\sqrt{\varepsilon_1}\cos\alpha_1 + \sqrt{\varepsilon_2}\cos\alpha_2} \qquad (5.172)$$

and

$$T_E = \frac{2\sqrt{\varepsilon_2}\cos\alpha_1}{\sqrt{\varepsilon_1}\cos\alpha_1 + \sqrt{\varepsilon_2}\cos\alpha_2}. \qquad (5.173)$$

The forms of eqns (5.167) and (5.172) suggest that for some incident angle the reflection coefficient might be zero. You can easily check that $\Gamma_H = 0$ when

$$\alpha_1 = \tan^{-1}\sqrt{(\varepsilon_2/\varepsilon_1)}. \qquad (5.174)$$

This value of α_1 is called the *Brewster angle.* A similar calculation will show that the condition $\Gamma_E = 0$ cannot be satisfied for any real α_1.

Is it possible to have zero transmission? You know that the answer is yes; you learned in school that *total reflection* occurs when $\varepsilon_2 < \varepsilon_1$ and α_1 is larger than a certain angle. The condition for the disappearance of the refracted wave is that $\sin\alpha_2 \geq 1$, i.e. α_2 is complex. Note that the derivations are unaffected. None of the mathematical operations required that α_2 was real hence the expressions for the field quantities are still valid. So we may formally write that

$$\cos\alpha_2 = \pm j\{(\varepsilon_1/\varepsilon_2)\sin^2\alpha_1 - 1\}^{\frac{1}{2}}, \qquad (5.175)$$

which leads to

$$T_H = \frac{2\sqrt{\varepsilon_2}\cos\alpha_1}{\sqrt{\varepsilon_2}\cos\alpha_1 \pm j\sqrt{(\varepsilon_1/\varepsilon_2)}(\varepsilon_1\sin^2\alpha_1 - \varepsilon_2)^{\frac{1}{2}}} \qquad (5.176)$$

and

$$T_E = \frac{2\sqrt{(\varepsilon_2)}\cos\alpha_1}{\sqrt{\varepsilon_1}\cos\alpha_1 \pm j(\varepsilon_2\sin^2\alpha_1 - \varepsilon_2)^{\frac{1}{2}}}. \qquad (5.177)$$

Oddly enough neither of the transmission coefficients show the slightest inclination to become zero. What's wrong? Let's first work out the reflection coefficients; we might be wiser afterwards. Substituting eqn (5.175) into eqns (5.167) and (5.172), we get

$$\Gamma_{\mathbf{H}}=\frac{\sqrt{\varepsilon_2}\cos\alpha_1\mp \mathrm{j}\sqrt{(\varepsilon_1/\varepsilon_2)}(\varepsilon_1\sin^2\alpha_1-\varepsilon_2)^{\frac{1}{2}}}{\sqrt{\varepsilon_2}\cos\alpha_1\pm \mathrm{j}\sqrt{(\varepsilon_1/\varepsilon_2)}(\varepsilon_1\sin^2\alpha_1-\varepsilon_2)^{\frac{1}{2}}} \tag{5.178}$$

and

$$\Gamma_{\mathbf{E}}=\frac{\sqrt{\varepsilon_1}\cos\alpha_1\mp \mathrm{j}(\varepsilon_1\sin^2\alpha_1-\varepsilon_2)^{\frac{1}{2}}}{\sqrt{\varepsilon_1}\cos\alpha_1\pm \mathrm{j}(\varepsilon_1\sin^2\alpha_1-\varepsilon_2)^{\frac{1}{2}}}. \tag{5.179}$$

It may be immediately seen that $|\Gamma_{\mathbf{H}}|^2=|\Gamma_{\mathbf{E}}|^2=1$, so our claim that the wave suffers total reflection appears to be upheld by the mathematics (for a proper proof one should really determine the Poynting vector). Can we have total reflection and some transmission at the same time? That is clearly impossible. Why aren't $T_{\mathbf{H}}$ and $T_{\mathbf{E}}$ equal then to zero? Because they are defined in terms of field quantities, and although no wave can propagate in medium 2, the field intensities can still be finite. How will the field appear there if propagation is not allowed? I am afraid our model is incapable of answering this question. Once we made the assumption that the time variation is $\exp \mathrm{j}\omega t$, we restricted our investigations to steady-state phenomena, and under steady-state conditions no electromagnetic power can move into medium 2. Of course we know the answer physically. When the plane wave was switched on, electromagnetic power moved into medium 2 and established the fields there or in other words stored up a certain amount of energy. Once the transients were over, no more power could move through the interface.

The variation of the field intensities in medium 2 can be easily obtained by substituting the imaginary value of $\cos\alpha_2$ into eqns (5.160) and (5.171). We get

$$\mathbf{E}_{\mathrm{T}}, \mathbf{H}_{\mathrm{T}}\sim\exp(-\mathrm{j}\beta_1 x\sin\alpha_1)\exp[-\beta_2\{(\varepsilon_1/\varepsilon_2)\sin^2\alpha_1-1\}^{\frac{1}{2}}z]. \tag{5.180}$$

Note that the value of $\cos\alpha_2$ has been taken with the negative sign so as the field decays in the direction of the positive z axis (an increasing exponential function is obviously impossible for energetic reasons).

5.16. Surface waves and dielectric waveguides

Let's start again with two dielectric media of different relative permittivity but take the interface at the plane $x=d/2$ and use the subscripts 1 and 2 for the upper and lower media respectively as shown in Fig. 5.20. Assume now that $\varepsilon_1<\varepsilon_2$ and that a plane wave is incident from medium 2 at such an angle that total reflection occurs (the magnetic field is chosen in the y direction).

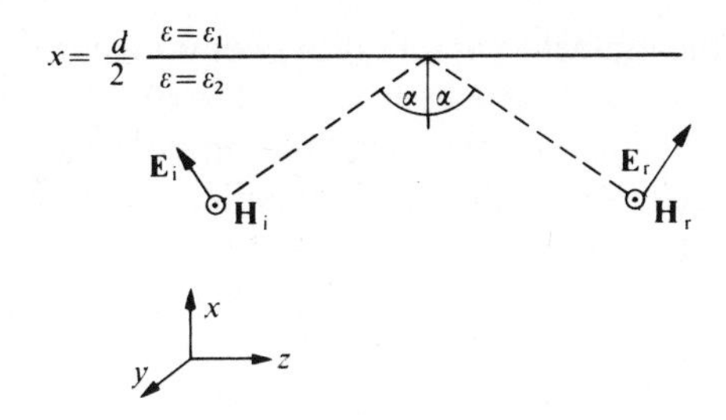

FIG. 5.20. Total reflection at the boundary of two dielectrics.

Then, as agreed in the previous section, the fields will decay exponentially away from the surface. So we could claim that the fields in the upper half-space stick in some sense to the $x = d/2$ surface.

Let us choose now a geometric configuration in which the material with relative permittivity ε_2 occupies only the $|x| < d/2$ region and fill the half-space $x < -d/2$ by a dielectric of relative permittivity ε_1, as shown in Fig. 5.21. To be more concrete we shall investigate the case when $\varepsilon_1 = \varepsilon_0$, i.e. the dielectric slab is bounded by vacuum on both sides.

It is no longer very meaningful to talk about a plane wave incident from medium 2 upon the interface $x = d/2$, but we shall retain that picture in a qualitative way. We shall claim that the wave incident at α_1 suffers total reflection at $x = d/2$ and that the same thing happens at the $x = -d/2$ interface. Hence the wave will propagate in the dielectric slab by means of total reflections. We know of course that this does not mean that all the fields are concentrated into the dielectric; there will be electric and magnetic fields in the surrounding vacuum as well but they will decay exponentially away from the two surfaces. There is no power flow from the dielectric into vacuum but there *is* power flow in the z direction in all three media. The mere fact that the wave propagating in the dielectric is totally reflected after each transit is not sufficient reason for claiming that a major part of the

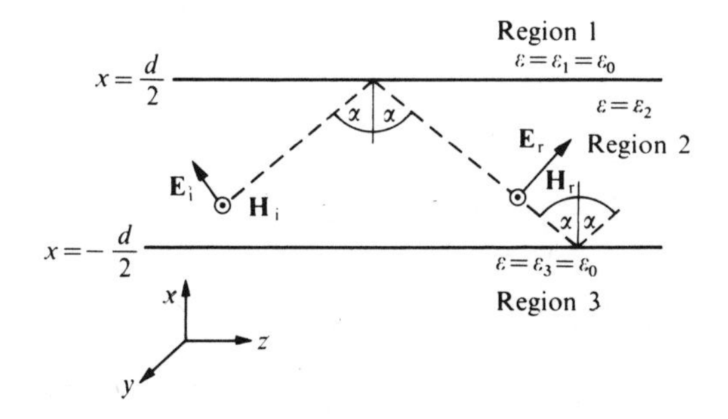

FIG. 5.21. Total reflections in a dielectric slab.

power is carried *in* the dielectric and that the fields die away quickly *outside* the dielectric. Nonetheless it appears to suggest such a possibility giving rise to the hope that we may use dielectrics for guiding electromagnetic waves.

It is time to conclude our qualitative considerations and start with the mathematics. We shall have to solve Maxwell's equations for the three media and match the solutions at the boundaries. With our choice of the fields in Figs 5.20 and 5.21 we shall look for a solution in which the fields have E_x, E_z, and H_y components and there is no change in the y direction $(\partial/\partial_y = 0)$.

We need to take again the curls

$$\nabla \times \mathbf{H} = \begin{vmatrix} \mathbf{i}_x & \mathbf{i}_y & \mathbf{i}_z \\ \dfrac{\partial}{\partial x} & 0 & \dfrac{\partial}{\partial z} \\ 0 & H_y & 0 \end{vmatrix} = -\mathbf{i}_x \frac{\partial H_y}{\partial z} + \mathbf{i}_z \frac{\partial H_y}{\partial x} \tag{5.181}$$

and

$$\nabla \times \mathbf{E} = \begin{vmatrix} \mathbf{i}_x & \mathbf{i}_y & \mathbf{i}_z \\ \dfrac{\partial}{\partial x} & 0 & \dfrac{\partial}{\partial z} \\ E_x & 0 & E_z \end{vmatrix} = \mathbf{i}_y \left(\frac{\partial E_x}{\partial z} - \frac{\partial E_z}{\partial x} \right), \tag{5.182}$$

leading to the scalar differential equations

$$-\frac{\partial H_y}{\partial z} = \mathrm{j}\omega\varepsilon E_x, \tag{5.183}$$

$$\frac{\partial H_y}{\partial x} = \mathrm{j}\omega\varepsilon E_z, \tag{5.184}$$

$$\frac{\partial E_x}{\partial z} - \frac{\partial E_z}{\partial x} = -\mathrm{j}\omega\mu H_y, \tag{5.185}$$

which after a few simple mathematical operations reduce to the single differential equation

$$\frac{\partial^2 H_y}{\partial x^2} + \frac{\partial^2 H_y}{\partial z^2} + \beta_{\mathrm{m}}^2 H_y = 0; \qquad \beta_{\mathrm{m}}^2 = \omega^2 \mu\varepsilon. \tag{5.186}$$

Assuming that the variables may be separated as

$$H_y(x, z) = X(x) Z(z), \tag{5.187}$$

we get the ordinary differential equations

$$\frac{\mathrm{d}^2 X}{\mathrm{d}x^2} + \xi^2 X = 0 \tag{5.188}$$

and

$$\frac{d^2Z}{dx^2}+\beta^2 Z=0, \tag{5.189}$$

where

$$\beta^2+\xi^2=\beta_m^2. \tag{5.190}$$

Since the fields need to be matched for all values of z at the $x=\pm d/2$ interfaces, the variation with z must be identical in all three regions, namely,

$$Z(z)=Z_f\,e^{-j\beta z}+Z_r\,e^{j\beta z}. \tag{5.191}$$

In order to save some work we can introduce here a few simplifications. Without any loss of generality we can take $Z_f=1$ and $Z_r=0$ (once we know the field distribution for the forward-travelling wave we know it also for the wave travelling in the reverse direction), and with some inevitable loss of generality (excluding in fact all the anti-symmetric solutions) we can restrict our investigations to solutions symmetric in x. Then in region 2 the solution for $X(x)$ is

$$X_2(x)=X_{20}\cos\xi_2 x; \qquad \xi_2^2=\omega^2\mu_0\varepsilon_2-\beta^2, \tag{5.192}$$

and in region 1

$$X_1(x)=X_{10}\exp\{-\chi_1(x-d/2)\}; \qquad -\chi_1^2=\xi_1^2=\omega^2\mu_0\varepsilon_0-\beta^2. \tag{5.193}$$

Note that χ_1 and ξ_2 are not independent of each other. In view of eqns (5.192) and (5.193) they are related as follows

$$\chi_1^2=\omega^2\mu_0(\varepsilon_2-\varepsilon_0)-\xi_2^2=\left(\frac{2\pi}{\lambda}\right)^2\left(\frac{\varepsilon_2}{\varepsilon_0}-1\right)-\xi_2^2. \tag{5.194}$$

We may now write for the magnetic fields

$$H_{y1}(x,z)=X_{10}\exp\{-\chi_1(x-d/2)\}\exp(-j\beta z) \qquad x\geqslant d/2 \tag{5.195}$$

and

$$H_{y2}(x,z)=X_{20}\cos\xi_2 x\exp(-j\beta z) \qquad |x|\leqslant d/2. \tag{5.196}$$

The components of the electric field may be obtained from eqns (5.183) and (5.184) in the form

$$E_{x1}=X_{10}\frac{\beta}{\omega\varepsilon_0}\exp\{-\chi_1(x-d/2)\}\exp(-j\beta z) \tag{5.197}$$

$$E_{z1}=-X_{10}\frac{\chi_1}{j\omega\varepsilon_0}\exp\{-\chi_1(x-d/2)\}\exp(-j\beta z) \tag{5.198}$$

and

$$E_{x2} = X_{20} \frac{\beta}{\omega\varepsilon_2} \cos \xi_2 x \, e^{-j\beta z} \tag{5.199}$$

$$E_{z2} = -X_{20} \frac{\xi_2}{j\omega\varepsilon_2} \sin \xi_2 x \, e^{-j\beta z}. \tag{5.200}$$

In order to satisfy the boundary conditions on the surface, we shall require the equality of the tangential components of the electric and magnetic fields

$$H_{y1}(d/2, z) = H_{y2}(d/2, z) \quad \text{and} \quad E_{z1}(d/2, z) = E_{z2}(d/2, z) \tag{5.201}$$

leading to the equations

$$X_{10} = X_{20} \cos(\xi_2 d/2) \tag{5.202}$$

and

$$\chi_1 X_{10} = \xi_2 X_{20} \sin(\xi_2 d/2). \tag{5.203}$$

Expressing X_{10}/X_{20} from eqn (5.202) and χ from eqn (5.194) we get the following relationship

$$(a^2 - u^2)^{\frac{1}{2}} = bu \tan u \tag{5.204}$$

where

$$u = \frac{\xi_2 d}{2}, \quad a = \frac{\pi d}{\lambda}\left(\frac{\varepsilon_2}{\varepsilon_0} - 1\right)^{\frac{1}{2}}, \quad b = \frac{\varepsilon_0}{\varepsilon_2}. \tag{5.205}$$

In contrast to the boundary value problems investigated in Section 5.12 we need to solve a transcendental equation to obtain the parameters χ_i and ξ_2. Nowadays one would use a computer to obtain all the solutions but let us be old-fashioned and try a graphical solution by plotting both sides of eqn (5.204). For $1/b = \varepsilon_2/\varepsilon_0 = 2{\cdot}26$ (corresponding to using polyethylene as the dielectric) and $a = 2$, 5, and 10, the curves are shown in Fig. 5.22. It is clear from the figure that the larger the value of a is the more solutions are possible.

We shall now, true to our habit, ignore the higher-order modes and restrict the investigations to the lowest one, obtained by the intersection of the circle with the first branch of the $bu \tan u$ curve. It may be seen from Fig. 5.22 that for all possible intersections $u < a$, hence from eqs (5.204) and (5.192))

$$\beta^2 = \omega^2 \mu_0 \varepsilon_2 - \left(\frac{2u}{d}\right)^2 > \omega^2 \mu_0 \varepsilon_2 - \left(\frac{2a}{d}\right)^2 = \omega^2 \mu_0 \varepsilon_0, \tag{5.206}$$

and

$$\chi_1^2 = \omega^2 \mu_0 (\varepsilon_2 - \varepsilon_0) - \left(\frac{2u}{d}\right)^2 > 0. \tag{5.207}$$

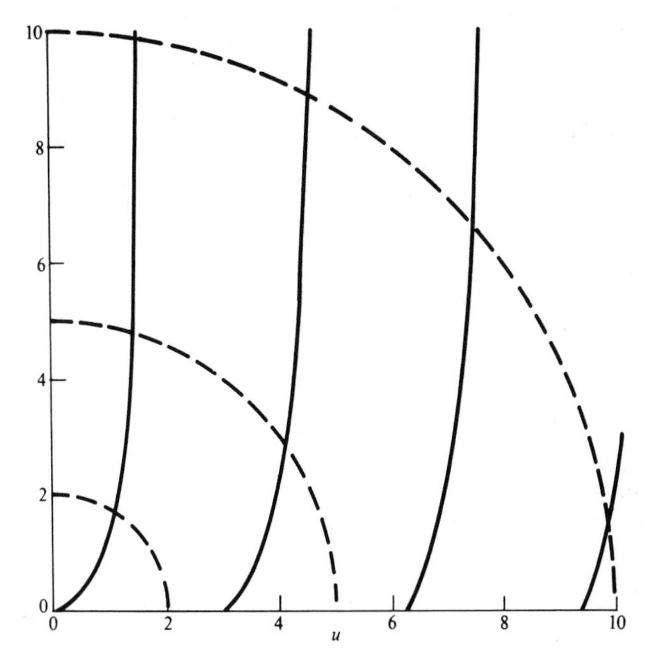

FIG. 5.22. ——— $u \tan u$, - - - - $(a^2 - u^2)^{\frac{1}{2}}$ for $a = 2, 5$, and 10 as functions of u.

Consequently, there is always a propagating solution however low the frequency is. This is a very interesting result suggesting that a dielectric slab is capable of guiding electromagnetic waves of any frequency.

Can you believe our good luck? Can Nature be that generous? My first impression would be the same as that of Laocoon when he came across the wooden horse, '*timeo Danaos et dona ferentes*', which translated into modern English says that 'Nature never offers boons to electrical engineers'.

Let's see how this waveguide will fare at low frequencies. We shall not choose an arbitrarily low frequency but one used in practice, $f = 50$ Hz, i.e. $\lambda = 6 \times 10^6$ m, and choose a reasonable thickness, say $d = 0{\cdot}02$ m. With the above choices we get $u \cong 1{\cdot}2 \times 10^{-8}$. This is a rather small value, making possible the approximate analytic solution of eqn (5.204), coming to

$$u^2 = a^2 - a^4 \tag{5.208}$$

whence

$$\chi_1 = \frac{2a^2}{d} = \frac{2\pi^2 d^2}{\lambda^2}\left(\frac{\varepsilon_2}{\varepsilon_0} - 1\right\} \tag{5.209}$$

and

$$\xi_2 \cong \frac{2a}{d} = \frac{2\pi}{\lambda}\left(\frac{\varepsilon_2}{\varepsilon_0} - 1\right)^{\frac{1}{2}}. \tag{5.210}$$

How will the field intensities vary? In region 2, E_{x2} varies as $\cos \xi_2 x$; at the boundary it reaches the value $\cos \xi_2 d/2 = \cos a$. Since a is so small there is practically no variation as a function of x in region 2. The variation in region 1 is given by the value of χ. The distance at which the fields reduce to $1/e$ of their value at the surface is $s = 1/\chi_1 \cong 1{\cdot}5 \times 10^{14}$ m, an astronomical distance.

The conclusion is that a dielectric slab of reasonable thickness will guide electromagnetic waves of any frequency, but the spread of the fields may turn out to be excessive. In the above example we overstepped the limit by many orders of magnitude.

For a second example we shall take again $d = 0{\cdot}02$ m, $\varepsilon_2 = 2{\cdot}26\varepsilon_0$, but take a wavelengh equal to the thickness, $\lambda = 0{\cdot}02$ m. Then $a = 3{\cdot}53$ and a graphical solution of eqn (5.204) yields $u = 1.39$. From eqns. (5.205) and (5.194) we get the values of the other parameters, $\xi_2 = 139\ \text{m}^{-1}$ and $\chi_1 = 324\ \text{m}^{-1}$. The normalized values of the electric field, E_x and E_z, as functions of x/d are plotted for this case in Fig. 5.23. It may be seen that there is strong decay beyond the dielectric boundary.

Is this type of waveguide used in practice? No, but a number of variations on the same theme *are* used. Before moving on to the other kinds let's pause

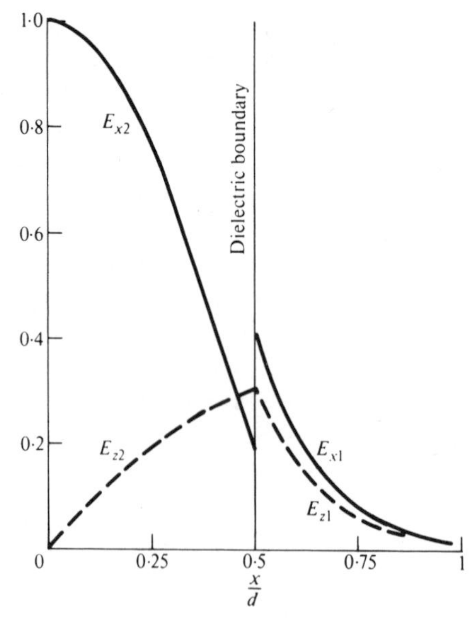

FIG. 5.23. The variation of the electric field in the transverse plane.

here for a moment and recapitulate what we have done so far. We have studied a two-dimensional geometry in which the magnetic field is in the transverse plane. We have solved Maxwell's equations subject to the boundary conditions. We have found a few higher-order modes of the symmetric variety but discussed the lowest mode only. Since the magnetic field is in the transverse plane these modes are called TM modes and the lowest one has the subscript 0 making it a TM_0 mode. There are of course TE modes as well which you could derive with similar ease.

Next we shall investigate the so-called *dielectric-coated waveguide.* It is a near relative of the dielectric slab, obtained by putting a metal plate into the $x = 0$ plane as shown in Fig. 5.24. Our previous solutions for the fields (eqns (5.195)–(5.205)) are still valid because the only new boundary condition that

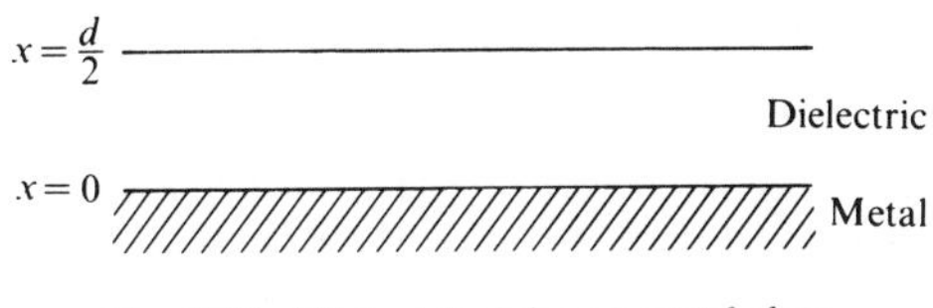

FIG 5.24. Dielectric slab on a metal plate.

E_z should vanish at $x = 0$ is automatically satisfied. Thus Fig. 5.23 gives correctly the field distribution for a half-wavelength-thick polyethylene layer resting on a metal plate.

We are now getting close to practical structures. Although the dielectric-coated plate is not used, the dielectric-coated metal (usually copper) wire does have applications. A typical cross-section is shown in Fig. 5.25(a), where the diameter of the dielectric is 3 times as large as the wire diameter.

How could we work out the field distributions? Well, we should perform the same kind of calculations as before but in cylindrical coordinates. The resulting differential equations, as you may guess, lead to Bessel functions of various kinds, which do bear some similarity to trigonometric and exponential functions. The number of zeros of the electric field in the dielectric will give the mode number, and again there will be strong (becoming exponential at large distances) decay outside the dielectric. There is now a single wire with a single surface that guides the electromagnetic wave so we are fully entitled to talk about surface waves and surface waveguides.

Finally, I want to talk about a newcomer to the family which might outshine them all. It rose to fame with the invention of lasers but has only recently acquired significance. I am talking about the *optical fibre*, the great white hope for optical communications. The wavelength is smaller but the principles are the same. In its simplest form it is just a dielectric cylinder as shown in Fig. 5.25(b). A more practical form is shown in Fig. 5.25(c) where

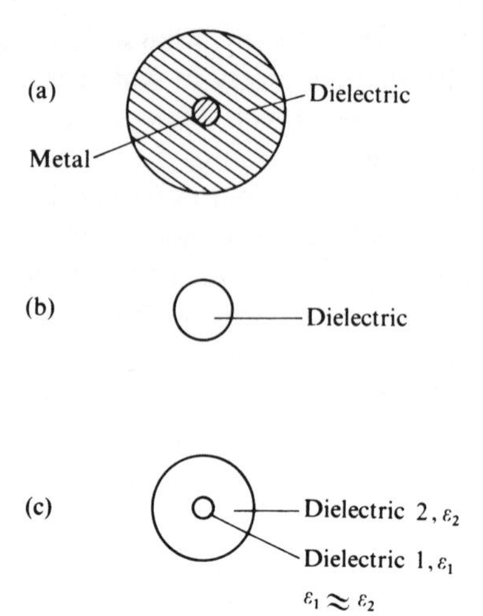

FIG. 5.25. (a) Dielectric-coated metal wire. (b) Dielectric cylinder. (c) Two concentric dielectric cylinders made of two materials of slightly different relative permittivity.

the two dielectrics have slightly different relative permittivities. The advantages of this latter realization are, (1) the outer diameter can be much larger than the wavelength and still only one mode propagates, and (2) the field strength is so small at the outer surface that the mechanical supports do not interfere with the propagation of the signal. I won't be able to prove the above statements here, you must believe for the moment that all these follow in a straightforward manner from the theory outlined.

Any other points of interest? Yes, we must consider losses. If we want a signal to travel a long distance, it must arrive at the other end of the line in a fair condition with a fair amplitude. How large attenuation is permissible? 20 db per km, say the experts. What sort of $\tan \delta$ does that imply? We may calculate the attenuation from eqn (5.105), which was derived in Section 5.11 for plane waves, but it is applicable for guided waves as well. For a length of 1 km the attenuation in decibels is given by the formula

$$A = 20 \log\{\exp(\tfrac{1}{2}\beta_m 1000 \tan \delta)\}$$

$$= 8680\pi \frac{\sqrt{\varepsilon'}}{\lambda} \tan \delta. \qquad (5.211)$$

For a typical glass fibre $\varepsilon' = 2{\cdot}25$, taking further an operating wavelength of $\lambda = 0{\cdot}8\ \mu\text{m}$ we obtain the result from the above equation that for $A = 20\ \text{db km}^{-1}$ we need $\tan \delta \cong 4 \times 10^{-10}$. In 1966 when the proposal was made the best material available exhibited $\tan \delta = 10^{-7}$. An improvement of more than 2 orders of magnitude was needed. The experts had grave doubts;

nonetheless the work started. By 1974 they managed to reduce the losses by 4 orders of magnitude, yielding an attenuation of 1 db per km, a remarkable performance. If sea-water was that transparent we could see the bottom of the sea at its deepest point.

5.17. Retarded potentials and the dipole antenna

Remember eqns (2.24) (p. 13) and (3.15) (p. 57) giving the scalar and vector potentials in terms of charges and currents; I shall write them down again in a slightly different form:

$$\phi(\mathbf{r}_2)=\frac{1}{4\pi\varepsilon}\int_\tau\frac{\rho(\mathbf{r}_1)}{|\mathbf{r}_1-\mathbf{r}_2|}\,\mathrm{d}\tau \tag{5.212}$$

and

$$\mathbf{A}(\mathbf{r}_2)=\frac{\mu}{4\pi}\int_\tau\frac{\mathbf{J}(\mathbf{r}_1)}{|\mathbf{r}_1-\mathbf{r}_2|}\,\mathrm{d}\tau. \tag{5.213}$$

They are valid for the electrostatic and magnetostatic cases respectively, ρ and $\mathbf{J}$ being independent of time. How will they modify when time variation is permitted? It is not very difficult to make an inspired guess.

In the static case the potential at $\mathbf{r}_2$ depends on the elementary charge $\rho\,\mathrm{d}\tau$ located at $\mathbf{r}_1$. Time does not come into consideration. It is assumed that the charge has always been at $\mathbf{r}_1$. The picture changes however when considering the build-up of the potential. We ask the following question: if we place a charge $\rho\,\mathrm{d}\tau$ into point $\mathbf{r}_1$ at time t_0, what is the potential at point $\mathbf{r}_2$ at the same t_0? Obviously zero. The information that a charge has been placed into point $\mathbf{r}_1$ has not yet reached point $\mathbf{r}_2$. It takes a finite time for the information to get there; an electromagnetic disturbance travels with the velocity of light. Thus, looking from $\mathbf{r}_2$, what matters is not what happens *now* at point $\mathbf{r}_1$ but what happened a time $|\mathbf{r}_2-\mathbf{r}_1|/c$ ago. Consequently, the potential at time t will depend on the value of ρ not at t but at a time $|\mathbf{r}_2-\mathbf{r}_1|/c$ earlier. The same applies of course to the relationship between current and the vector potential. Hence our formulae will take the following form for the time-varying case,

$$\phi(\mathbf{r}_2,t)=\frac{1}{4\pi\varepsilon_0}\int_\tau\frac{\rho(\mathbf{r}_1,t-|\mathbf{r}_2-\mathbf{r}_1|/c)}{|\mathbf{r}_2-\mathbf{r}_1|}\,\mathrm{d}\tau \tag{5.214}$$

and

$$\mathbf{A}(\mathbf{r}_2,t)=\frac{\mu_0}{4\pi}\int_\tau\frac{\mathbf{J}(\mathbf{r}_1,t-|\mathbf{r}_2-\mathbf{r}_1|/c)}{|\mathbf{r}_2-\mathbf{r}_1|}\,\mathrm{d}\tau. \tag{5.215}$$

This is what may be called a *heuristic* derivation. We argued from the static case instead of solving the time-varying equations. In spite of the lack of rigour the above equations happen to be correct. A rigorous derivation is fairly difficult (measured by the standard of this course) but if you are interested you can look it up in a number of textbooks.† The potentials of eqns (5.214) and (5.215) are usually referred to as *retarded* potentials, because in order to obtain them we need the retarded values (those at an earlier time) of the charges and currents.

We shall now consider the simplest possible case when

$$\mathbf{J}\,\mathrm{d}\tau = \frac{I}{S_0} S_0\,\mathrm{d}l\,\mathbf{i}_z = I\,\mathrm{d}l\,\mathbf{i}_z, \tag{5.216}$$

where $\mathrm{d}l$ is the longitudinal extent of the current element pointing in the z direction, and the cross-section S_0 is assumed to be infinitesimally small. This current element is placed at the centre of the coordinate system as shown in Fig. 5.26. Where does the current flow to? Nowhere. We haven't

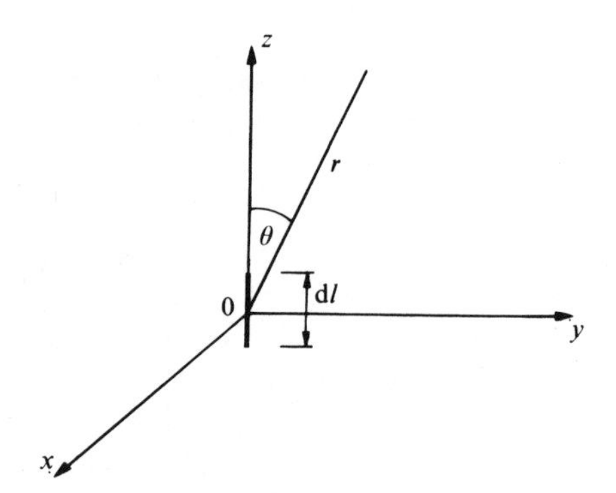

FIG. 5.26. An electric dipole.

got a closed circuit in the low-frequency sense, and there isn't much point either in invoking the displacement current in order to save the concept of a closed circuit. What is then the correct way of looking at the current element? I am not sure that there *is* a correct way, but I can tell you how the subject is usually introduced. Imagine an elementary dipole with two charges of opposite sign sitting close to each other. Then imagine that both charges vary sinusoidally as a function of time, implying that continuous charge transfer takes place, i.e. a current flows within the range $\mathrm{d}l$. If you want a somewhat more realistic model (in the sense that it is nearer to the practical realization of a current element) imagine a short piece of wire in which charges can move up and down. For a time $T/2$ they all move

† See, for example, D. S. JONES (1964) *The theory of electromagnetism*, Section 1.16. Pergamon Press, Oxford.

upwards, and in the next $T/2$ interval they all move downwards, creating thereby an alternating current.

Let's now go over to the mathematics. Since $\mathrm{d}l$ is an elementary distance and since we are interested in working out the fields for macroscopic distances large in comparison with $\mathrm{d}l$ we can assume that I is independent of the spatial coordinates. As a function of time we need to retard the current that is to take its value at a time $t-r/c$ earlier. If we consider (as usual) a harmonic time variation then the current is of the form

$$I(t-r/c)=I_0\exp\{\mathrm{j}\omega(t-r/c)\}=I_0\exp\{\mathrm{j}(\omega t-\beta_0 r)\}, \tag{5.217}$$

or after dropping (as usual) the $\exp \mathrm{j}\omega t$ factor we are left with

$$I=I_0\exp(-\mathrm{j}\beta_0 r). \tag{5.218}$$

Substituting eqns (5.216) and (5.218) into eqn (5.213), dropping the integration sign, and noting that $|\mathbf{r}_2-\mathbf{r}_1|$ is now equal to the distance r in spherical coordinates (Fig. 5.26), we get for the z (and only) component of the vector potential

$$A_z(r,\theta)=\frac{\mu_0 I\,\mathrm{d}l}{4\pi}\,\frac{\exp(-\mathrm{j}\beta_0 r)}{r}. \tag{5.219}$$

Before working out the field quantities there is one more technical thing to do; the A_z component of the vector potential needs to be expressed in spherical coordinates. As follows from Fig. 5.27, we get

$$A_r=A_z\cos\theta \quad \text{and} \quad A_\theta=-A_z\sin\theta, \tag{5.220}$$

leading to the equation for the magnetic field

$$\mathbf{H}=\frac{1}{\mu_0}(\nabla\times\mathbf{A})=\frac{1}{\mu_0 r^2\sin\theta}\begin{vmatrix}\mathbf{i}_r & r\mathbf{i}_\theta & r\sin\theta\,\mathbf{i}_\varphi\\ \dfrac{\partial}{\partial r} & \dfrac{\partial}{\partial\theta} & 0\\ A_z\cos\theta & -rA_z\sin\theta & 0\end{vmatrix}. \tag{5.221a}$$

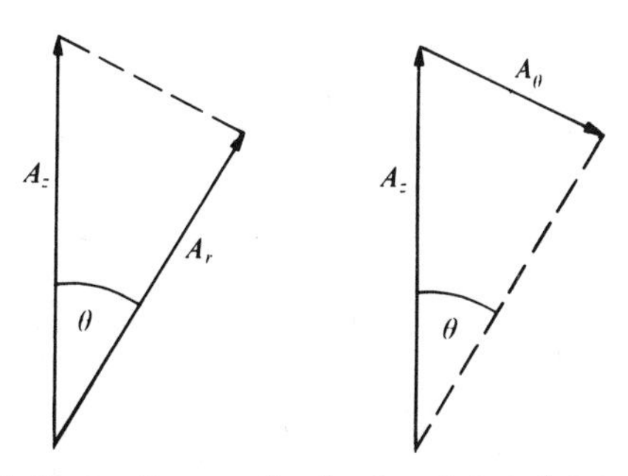

FIG. 5.27. Vector diagrams showing the relation of A_z to A_r and A_θ.

As may be seen there will be an H_φ component only, which after a few mathematical operations may be written in the form

$$H_\varphi = \frac{I_0 \mathrm{d}l}{4\pi} \frac{\exp(-\mathrm{j}\beta_0 r)}{r}\left(\mathrm{j}\beta_0 + \frac{1}{r}\right) \sin\theta. \tag{5.221b}$$

The electric field may now be obtained from

$$\mathbf{E} = \frac{1}{\mathrm{j}\omega\varepsilon_0} \nabla \times \mathbf{H}, \tag{5.222}$$

yielding

$$\left.\begin{aligned} E_r &= -\mathrm{j}\frac{I_0\, \mathrm{d}l}{2\pi\omega\varepsilon} \frac{\exp(-\mathrm{j}\beta_0 r)}{r^2}\left(\mathrm{j}\beta_0 + \frac{1}{r}\right) \cos\theta, \\ E_\theta &= -\mathrm{j}\frac{I_0\, \mathrm{d}l}{4\pi\omega\varepsilon} \frac{\exp(-\mathrm{j}\beta_0 r)}{r}\left(-\beta_0^2 + \frac{\mathrm{j}\beta_0}{r} + \frac{1}{r^2}\right) \sin\theta, \\ E_\varphi &= 0. \end{aligned}\right\} \tag{5.223}$$

First of all let's try to establish some relationship with the static dipole. There we introduced the electric dipole moment with the relation $p_e = qd$. If we go back to our picture of the current element as due to the motion of charges infinitesimally close to each other then the length of our current element $\mathrm{d}l$ is the same as d, the distance between the charges, leading to the relationship

$$I_0\, \mathrm{d}l = \mathrm{j}\omega q d = \mathrm{j}\omega p_e. \tag{5.224}$$

Substituting the above expression into eqn (5.223) and assuming that $\beta_0 r \to 0$ we find that

$$E_r = \frac{p_e}{2\pi\varepsilon_0} \frac{\cos\theta}{r^3} \quad \text{and} \quad E_\theta = \frac{p_e}{4\pi\varepsilon_0} \frac{\sin\theta}{r^3}, \tag{5.225}$$

in agreement with eqn (2.33). Thus our eqn (5.223) contains the electrostatic dipole as a special case.

Let us investigate now the other limit. The condition $\beta_0 r \to \infty$ leads to

$$\left.\begin{aligned} H_\varphi &= \mathrm{j}\frac{\beta_0 I_0\, \mathrm{d}l}{4\pi} \frac{\exp(-\mathrm{j}\beta_0 r)}{r} \sin\theta, \\ E_r &= 0, \\ E_\theta &= \mathrm{j}\frac{\beta_0^2 I_0\, \mathrm{d}l}{4\pi\omega\varepsilon} \frac{\exp(-\mathrm{j}\beta_0 r)}{r} \sin\theta. \end{aligned}\right\} \tag{5.226}$$

We shall say more about the above expression later; let's first calculate the Poynting vector.

$$\mathbf{P}_d = \tfrac{1}{2}\,\mathrm{Re}(E_\theta \mathbf{i}_\theta \times H_\varphi^* \mathbf{i}_\varphi)$$
$$= \frac{1}{2} Z_0 \left(\frac{\beta_0 I_0\, \mathrm{d}l}{4\pi r}\right)^2 \sin^2\theta\, \mathbf{i}_r. \tag{5.227}$$

The conclusion is that there is power flow *radially outwards.* You should not underestimate the significance of this conclusion; it is radically different from anything we have come across so far. There is a current element at the origin, charges move up and down and, lo and behold, power flows out in every (except in the axial) direction.

Note that this is a solution very much unlike those for plane waves or guided waves. There we showed that a wave may propagate in free space or that we have means of persuading waves to propagate in some prescribed direction, but we couldn't say anything about the origin of the wave, how it came into existence. Now we have an example. Charges moving up and down produce electromagnetic waves. Is this a practical way of producing electromagnetic waves? Yes, it is. In general, the electromagnetic wave moving outwards is referred to as *radiation*, and the structure that produces it is called an *antenna.*† Our piece of wire, the time-varying equivalent of the static dipole, is called a *dipole antenna.*

What sort of wave is produced by a dipole antenna? Well, it moves radially outwards so we could call it a spherical wave. Is a spherical wave much different from a plane wave? Not in our case. As may be seen from eqn (5.226) the electric and magnetic fields are perpendicular to each other and

$$E_\theta / H_\varphi = Z_0, \tag{5.228}$$

the free-space impedance; so there are lots of similarities. The only major difference is perhaps the presence of sin θ in eqn (5.226). For plane waves the field quantities depended on one coordinate only (the direction of propagation), in our spherical wave there is variation with two coordinates, r and θ.

Is it reasonable that radiated power decays as $1/r^2$? Once we admit that electromagnetic power moves radially outwards, the power density *must* decay as $1/r^2$. This may be seen by taking an elementary solid angle in which the radiating power $P\mathrm{d}\Omega$ is constant (independent of angle). At a distance r away this power will cross a surface $\mathrm{d}S = r^2\,\mathrm{d}\Omega$, hence the power density is

$$P_d = \frac{P\,\mathrm{d}\Omega}{\mathrm{d}S} = \frac{P}{r^2}, \tag{5.229}$$

inversely proportional to r^2.

† From the Greek *anateino* = to stretch out. The word antenna was used well before the advent of electronics for describing the sensory organs of certain insects.

I shall not be able to describe antennas in any detail, but I would like to give a brief introduction to the subject and that will make it necessary to talk about three characteristics of antennas, namely radiation pattern, gain, and radiation resistance.

As you may notice from eqn (5.226), H_φ and E_θ depend on the angle θ. In general, the angular dependence of the field quantities is referred to as the radiation pattern which we shall denote by $g(\theta, \varphi)$. It is customary to introduce some normalization so that $g(\theta_m, \varphi_m) = 1$, where θ_m, φ_m represent the direction in which the radiation is maximum. The radiation pattern of the

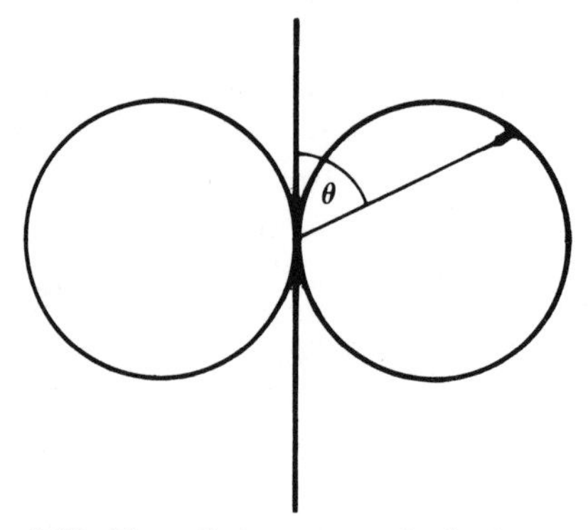

FIG. 5.28. The radiation pattern of a dipole antenna.

dipole antenna (plotted in Fig. 5.28 in polar coordinates) is independent of φ and depends on θ as

$$g(\theta) = \sin\theta, \qquad (5.230)$$

the direction of maximum radiation being $\theta_m = 90°$. The gain is defined as the power density in the direction of maximum radiation produced by the antenna under consideration, divided by the power density produced by a hypothetical isotropic (radiating equally in all directions) antenna. As the name implies, our measure is how much we *gain* in power density by using a directional antenna.

In order to determine the gain of a dipole antenna we need to do two more things, (1) work out the power density produced by an isotropic antenna, (2) work out the total power radiated by a dipole antenna. The first problem is a fairly easy one to solve. If we feed a power P into an isotropic antenna, it will be uniformly distributed in all directions, hence the power density at a distance r from the antenna will be

$$P_d = \frac{P}{4\pi r^2}. \qquad (5.231)$$

The total power radiated by the dipole antenna may be obtained by integrating the power density (eqn (5.227)) over a spherical surface

$$P = \int P_d \, dS = \frac{1}{2} Z_0 \left(\frac{\beta_0 I_0 \, dl}{4\pi r} \right)^2 \int_0^{2\pi} \int_0^{\pi} \sin^2\theta r^2 \sin\theta \, d\theta \, d\varphi$$

$$= \frac{1}{12\pi} Z_0 (\beta_0 I_0 \, dl)^2. \quad (5.232)$$

We are now in a position to calculate the gain. From eqns (5.227) and (5.232) the power density produced by the dipole antenna in the direction $\theta_m = 90°$ is

$$P_d = \frac{3P}{8\pi r^2}. \quad (5.233)$$

Feeding the same power P into an isotropic antenna, it would produce a power density $P/4\pi r^2$, as agreed before; hence the gain of a dipole antenna is

$$G = \frac{3P/8\pi r^2}{P/4\pi r^2} = \frac{3}{2}. \quad (5.234)$$

It is a fact of life that the properties of antennas can only be unravelled with the aid of electromagnetic theory, but there is another fact of life, namely that electrical engineers who know field theory need to communicate with electrical engineers who have only very vague ideas about fields. Thus the concept of radiation resistance was introduced some time ago, mainly for the benefit of circuit engineers. The logic as as follows. An antenna radiates out power, hence power is lost from the system. Lost power may be represented by a resistance. Consequently, we may define radiation resistance by the relation

$$\tfrac{1}{2} R_{rad} I_0^2 = P. \quad (5.235)$$

For a dipole antenna (from eqn (5.232))

$$R_{rad} = \frac{2\pi}{3} Z_0 \left(\frac{dl}{\lambda} \right)^2. \quad (5.236)$$

We have so far always emphasized that we have been concerned with a current element, with an infinitesimal length of wire over which the current is constant. What happens when the wire length is comparable with the wavelength? What is, for example, the current distribution for a half-wave dipole?

A rigorous answer† is beyond the scope of the present course and is, in fact, beyond the scope of most antenna courses. Nevertheless, one can give a qualitative argument that will suffice for our purpose.

We have often come across the travelling wave described mathematically as $\exp j(\omega t - \beta z)$. Let us assume that our wire is excited symmetrically (Fig. 5.29(a)) by some source of a.c. power, and that such waves will travel along the wire in both directions (Fig. 5.29(b)). The tip of the wire represents a

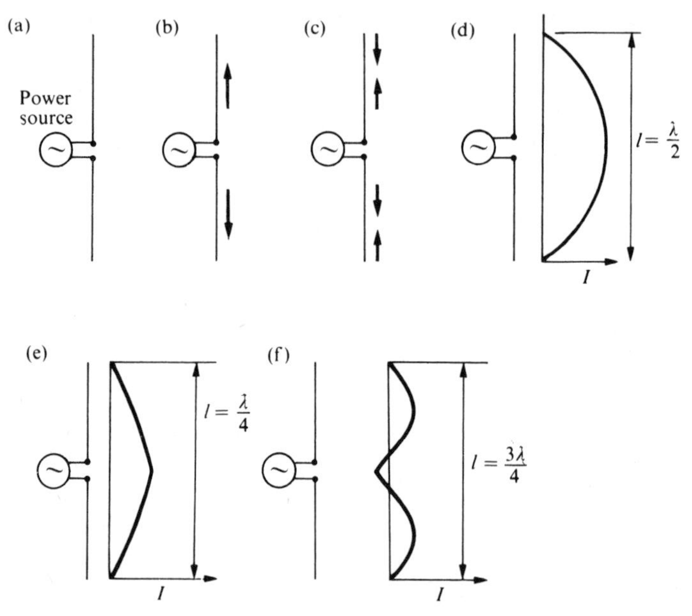

FIG. 5.29. (a) Dipole antenna excited by an external source. (b) Waves move outwards. (c) Waves move in both directions owing to reflections at the end of the wire. (d) $\lambda/2$, (e) $\lambda/4$, and (f) $3\lambda/4$, lengths and the current distribution along the dipole.

large discontinuity. To a first-order approximation the wave is completely reflected (Fig. 5.29(c)) producing a sinusoidal standing wave as shown in Fig. 5.29(d). For a length of half-wavelength we have a half-sinusoidal. If the wire is say, a quarter wave or three-quarter wave long then the standing wave will be constituted by the corresponding sections of the sine function (Figs 5.29(e) and (f)).

† If one wants to be rigorous one should first describe the wire as a cylinder of finite length and of finite diameter made of a material of finite conductivity. Then one should find such a solution of Maxwell's equations which satisfies the boundary conditions on the surface of the cylinder. A very messy problem, even if infinite conductivity is assumed. The trouble comes from the finite length of the cylinder. There is no coordinate system in which a constant-coordinate surface coincides with that of a finite cylinder (remember there are top and bottom surfaces as well).

One last question. How can we determine the electric and magnetic fields if the current is not constant? Use the principle of superposition. Assume that the current is constant in each small section of the wire and then sum the contributions from each section. This means that the vector potential will be given (instead of eqn (5.219)) by

$$A_z(r_0, \theta) = \frac{\mu_0}{4\pi} \int_{-l/2}^{l/2} \frac{I(z)\exp(-j\beta_0 r)}{r}\, dz, \tag{5.237}$$

where r_0, θ are the coordinates of the point P in which the vector potential is to be determined, l is the length of the wire, r is the distance from the current element to the point P (Fig. 5.30), and $I(z)$ is the assumed variation of

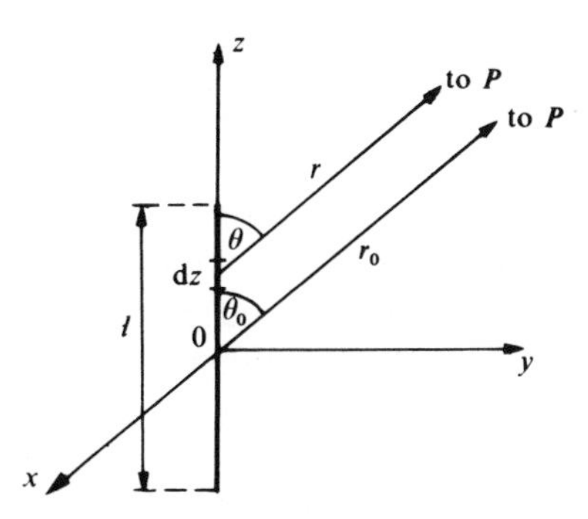

FIG. 5.30. A dipole of length l. Radiation pattern is calculated by summing the contributions from each current element.

current along the wire. If $r_0 \gg l$ (i.e. the radiation is to be determined far away from the antenna) the vector potential—and with its aid the magnetic and electric fields—may be fairly easily determined (Example 5.27).

5.18. Geometrical optics

What has geometrical optics got to do with electromagnetic theory? Well, light is an electromagnetic phenomenon, and geometrical optics is concerned with the behaviour of light, so all geometrical optics should follow from Maxwell's equations. In fact, we have already derived a number of relationships which could be regarded as foundation stones of geometrical optics. For example, plane waves (plane wavefront) or spherical waves (spherical wavefront) correspond to rays moving parallel or diverging from a point source. We have managed to identify the refractive index with the square root of the relative permittivity and derived the laws of reflection and refraction at boundaries. Is there anything more to do? Yes, geometrical optics means a little more. It is not concerned with infinite wavefronts; it claims that the wavefront at *every point* may be regarded as that of a local plane wave and that rays and wavefronts are always perpendicular to each other even under conditions when the refractive index is space dependent and the rays are not straight. When do we have curved rays? At sunset for example. You probably know that we still see the sun when it is already beyond the horizon. So we should look at Maxwell's equations again, allowing this time the variation of permittivity with the spatial coordinates.

We shall start, as so many times before, with the equations

$$\nabla \times \mathbf{H} = \mathrm{j}\omega\varepsilon \mathbf{E} \tag{5.238}$$

and

$$\nabla \times \mathbf{E} = -\mathrm{j}\omega\mu \mathbf{H} \tag{5.239}$$

but assume this time the solution in the form

$$\mathbf{E} = \mathbf{e}\exp(-\mathrm{j}\beta_0 L), \qquad \beta_0 = \omega\sqrt{(\mu_0\varepsilon_0)}, \tag{5.240}$$

and

$$\mathbf{H} = \mathbf{h}\exp(-\mathrm{j}\beta_0 L), \tag{5.241}$$

where **e**, **h**, and L are real functions of the spatial coordinates. Substituting now eqns (5.240) and (5.241) into (5.238) and using the vector relationship (A.3) we get

$$\mathbf{e} = \frac{1}{\mathrm{j}\omega\varepsilon}\{-\mathrm{j}\beta_0(\nabla L \times \mathbf{h}) + \nabla \times \mathbf{h}\}. \tag{5.242}$$

And now comes the crucial approximation. Geometrical optics is valid when all the dimensions of interest are large in comparison with the wavelength of the electromagnetic wave, or in other words when the wavelength tends to zero or when the frequency tends to infinity. Hence we are going to neglect the second term in the braces of eqn (5.242), leading to the fairly simple expression

$$\mathbf{e} = -\frac{\beta_0}{\omega\varepsilon}\nabla L \times \mathbf{h}. \tag{5.243}$$

Entirely analogous substitutions into eqn (5.239) and application of the $\omega \to \infty$ condition yields

$$\mathbf{h} = \frac{\beta_0}{\omega\mu_0}\nabla L \times \mathbf{e}. \tag{5.244}$$

Eliminating **h** from eqn (5.243) with the aid of eqn (5.244), we obtain a vector equation

$$\mathbf{e} = -\frac{\beta_0^2}{\omega^2\mu_0\varepsilon}\{\nabla L(\mathbf{e}\nabla L) - \mathbf{e}(\nabla L)^2\}, \tag{5.245}$$

which is satisfied if

$$\mathbf{e}\nabla L = 0 \tag{5.246}$$

and

$$(\nabla L)^2 = \frac{\omega^2 \mu_0 \varepsilon}{\beta_0^2} = \frac{\varepsilon}{\varepsilon_0} = n^2. \tag{5.247}$$

A similar calculation for **h** yields the condition

$$\mathbf{h}\nabla L = 0. \tag{5.248}$$

What are L and ∇L? We have introduced L in eqn (5.240) as a real function without saying anything about its physical significance. It is not difficult to see that the equation

$$L(x, y, z) = L_0 = \text{constant} \tag{5.249}$$

gives the surfaces over which the phase is constant. We may call these equiphase surfaces or (in the language of geometrical optics) *wavefronts*. ∇L gives then a vector that is normal to the wavefront. If we define now rays as the orthogonal trajectories of wavefronts it is clear then that the vector ∇L gives the direction of a ray in every point in space.

The mathematical conditions (5.246) and (5.248) state that the electric and magnetic fields are always perpendicular to the rays, and it follows from eqns (5.243) and (5.244) that **e** and **h** are perpendicular to each other. Hence power flows in the direction of the rays, and the expression for the Poynting vector is as follows

$$\begin{aligned}\mathbf{P}_{\mathrm{d}} &= \tfrac{1}{2}\,\mathrm{Re}(\mathbf{E}\times\mathbf{H}^*) = \tfrac{1}{2}(\mathbf{e}\times\mathbf{h}) = \frac{\beta_0}{2\omega\mu_0}\mathbf{e}\times(\nabla L\times\mathbf{e})\\ &= \frac{1}{2Z_{\mathrm{m}}}|\mathbf{e}|^2\mathbf{i}_{\mathrm{p}} = \frac{1}{2Z_{\mathrm{m}}}|\mathbf{E}|^2\mathbf{i}_{\mathrm{p}},\end{aligned} \tag{5.250}$$

where

$$\mathbf{i}_{\mathrm{p}} = \frac{1}{n}\nabla L \tag{5.251}$$

is the unit vector in the direction of propagation and Z_{m} is the impedance of the medium. In any practical case if the variation of the refractive index is given, we can find the rays from the differential equations derived.

Is the use of geometrical optics restricted to the optical region? No, not at all. The condition that the dimensions of interest should be large in comparison with the wavelength can be met at much larger wavelengths. Reflector antennas working in the microwave region are invariably designed with the aid of the ray concept.

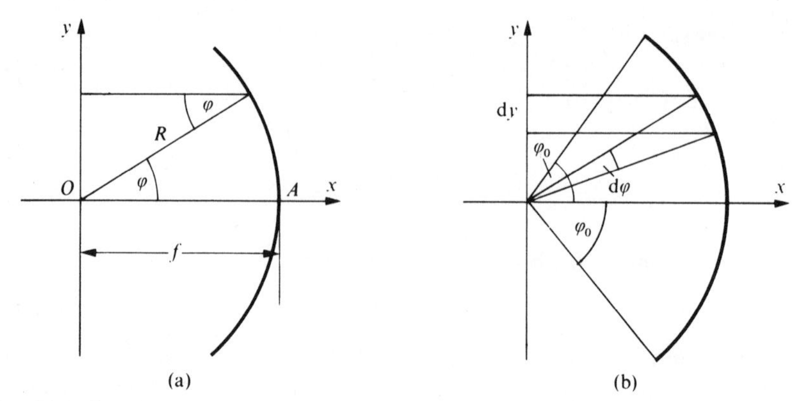

FIG. 5.31. Cylindrical wavefront is transformed into a plane wavefront by a parabolic reflector.

Let us quickly design a reflector that will transform a cylindrical wavefront into a plane wavefront. This is a two-dimensional problem. The source is situated at point O (Fig. 5.31(a)), and for simplicity we shall assume that its radiation is independent of the angle φ. The aim is that after the reflection all the rays should be parallel to the x axis. An alternative expression of the same condition is that the paths of all rays reaching the y axis should be the same (hence phases identical).

The mathematical formulation of the above condition is very simple. The ray reflected straight back at A has a path of $2f$, whereas the ray leaving at an angle φ will have a path $R+R\cos\varphi$. Hence the equation for the reflector shape is

$$R=\frac{2f}{1+\cos\varphi}, \tag{5.252}$$

which is the equation of a parabola in polar coordinates. Thus in order to transform a cylindrical wavefront into a plane wavefront we need a cylindrical parabola as a reflector. Have we produced a plane wave? We have produced a plane wavefront but not a plane wave. In a plane wave the electric and magnetic fields are constant in a transverse plane, but that is not the case for this reflected wave. The power (and intensity of the electric field) will vary as a function of y as may be seen from the following considerations.

Power, as we agreed, flows along a ray, thus the amount of power between two rays cannot vary. Consequently, the power carried within the elementary angle $\mathrm{d}\varphi$ must appear within the elementary interval $\mathrm{d}y$, as shown in Fig. 5.31(b). If we assume that the total amount of power radiated within the angles $-\varphi_0$ and φ_0 is P, and we denote by P_y the power per unit length crossing the y axis then

$$P_y\,\mathrm{d}y=P\,\mathrm{d}\varphi. \tag{5.253}$$

Noting that

$$y = R \sin \varphi = 2f \tan \frac{\varphi}{2}, \tag{5.254}$$

we can express $d\varphi$ in terms of dy yielding

$$P_y \, dy = \frac{4fP}{4f^2 + y^2} \, dy. \tag{5.255}$$

For $\varphi_0 = 75°$ the distribution of the electric field (proportional to the square root of power density) is plotted (curve (a) in Fig. 5.32). It may be seen that

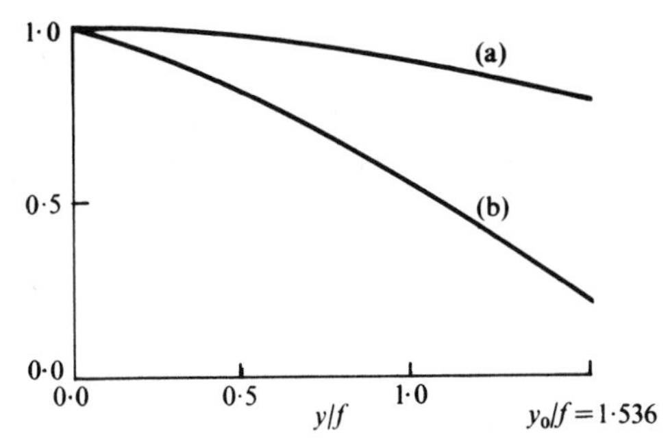

FIG. 5.32. The normalized distribution of electric field for (a) $g(\varphi) = 1$, and (b) $g(\varphi) = \cos \varphi$.

the electric field decays towards the edge of the reflector, although the source (usually called the primary feed) at O was assumed to radiate equally in all directions. If we take a more realistic primary feed pattern which varies as $g(\varphi) = \cos \varphi$, the decay towards the edge is more pronounced as shown in curve (b).

5.19. Huygens' principle and aperture antennas

Let us start with a concrete example and take a plane wave incident perpendicularly upon a rectangular aperture cut into an infinite metal plate (Fig. 5.33) located at $z = 0$. Can we find the field at an arbitrary point of the half-space $z > 0$? The equations we have derived up to now are of little use in solving this problem. On the basis of geometrical optics we could predict that the plane wave is reflected by the metal plate everywhere except at the aperture where it will carry on unhindered.

Taking the incident wave in the form

$$\mathbf{E} = E_0 \mathbf{i}_y \exp(-j\beta_0 z) \tag{5.256}$$

and the coordinates of the aperture as $(-a/2, a/2)$ in the x direction and

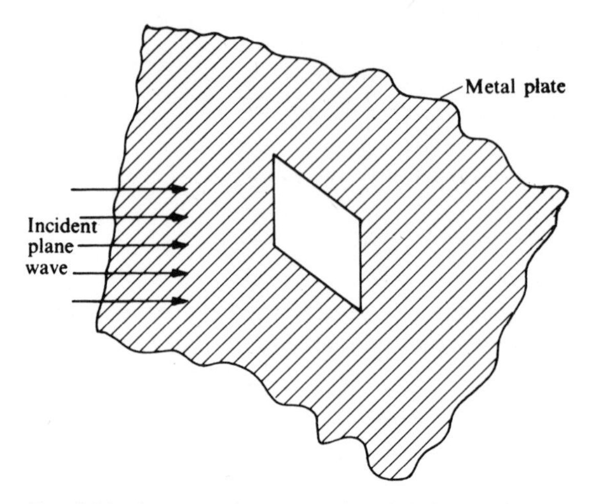

FIG. 5.33. A rectangular aperture in an infinite metal plate.

$(-b/2, b/2)$ in the y direction; the prediction of geometrical optics is that

$$\begin{aligned} &\mathbf{E}=0 && |x|>a/2, |y|>b/2, \\ &\mathbf{E}=E_0\mathbf{i}_y \exp(-\mathrm{j}\beta_0 z) && |x|<a/2, |y|<b/2. \end{aligned} \tag{5.257}$$

This may be true in the immediate vicinity of the aperture but can hardly be valid further on. We know that there is such a phenomenon as diffraction; light is known to be capable of going round a corner. How can we obtain this diffracted field? Let us recall our solution for the dipole antenna. We have found that the electric field had only an E_θ component, and it could be written (apart from the sin θ factor) as

$$E_\theta \sim \frac{I_0\,\mathrm{d}l}{r}\exp(-\mathrm{j}\beta_0 r). \tag{5.258}$$

What happens if instead of a current element we have a current distribution $I(z)$? Then, we could argue, each element emits a spherical wave which need to be added in phase. For the geometry of Fig. 5.30 the electric field may be written as

$$E_\theta \sim \int I(z)\frac{\exp(-\mathrm{j}\beta_0 r)}{r}\,\mathrm{d}z. \tag{5.259}$$

What is so special about current elements? If charges moving up and down can emit spherical waves, couldn't something else varying fast also give rise to spherical waves? The answer is yes; a time-varying electric field that exists over a surface element can just as well be regarded as a source of radiation.†

Thus purely on the basis of the above analogy we may write the electric field far away from the aperture as

$$E_\theta \sim \int E_A \frac{\exp(-j\beta_0 r)}{r} \, dS, \tag{5.260}$$

where E_A is the electric field in the aperture (not necessarily constant). Eqn (5.260) is nothing else but the mathematical formulation of Huygens' principle you met in school. We choose a surface (the aperture in the metal plane in this case) through which time-varying electric fields are established and then claim that each element of this surface gives rise to spherical waves to be added in the right phase.

I am again guilty of presenting an heuristic approach. If Huygens' principle does follow from Maxwell's equations why didn't I show it in clear mathematical terms instead of relying on physical arguments? It is indeed possible to derive a formula from Maxwell's equation in which the electric and magnetic fields at an arbitrary point in space are related to the electric and magnetic fields over a closed surface but the derivation is lengthy and difficult. It requires mathematical rigour (one of the places where I admit rigour is necessary) and a clear view about conditions reigning at infinity. I have seen strong-willed postgraduate students reduced to nervous wrecks after their third attempt at the derivation. I am afraid the formula is not for undergraduate consumption, not in this nor in any other course. And besides, the formula turns out to be much more complicated than eqn (5.260). Owing to the vectorial character of the fields a number of extra factors come in. Luckily, for configurations of practical importance to engineers, when the aperture is large in comparison with the wavelength, a number of further approximations may be made and the formula may be reduced to that of eqn (5.260) providing at the same time the missing constants. We shall gratefully accept the constants with which our eqn (5.260) modifies to

$$E_\theta = -j\frac{\beta_0}{2\pi} \int E_A \frac{\exp(-j\beta_0 r)}{r} \, dS. \tag{5.261}$$

We shall now evaluate eqn (5.261) under the assumption that the distance at which the electric field is to be determined is much larger than the dimensions of the aperture. The notation is shown in detail in Fig. 5.34 for the rectangular aperture assumed. We take an arbitrary surface element dS

† Some care needs to be exercised here. In order to produce an electric field some currents and charges are necessary (quantum-mechanical emitters are also possible but we shan't be able to discuss them in this course) but having produced a well-defined electric field over a surface, we may now regard each surface element as a radiator.

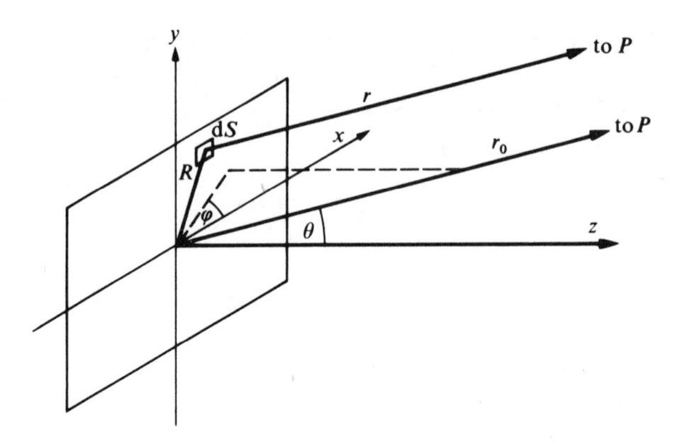

FIG. 5.34. The contribution to the far field of a surface element.

located at x, y over which the field intensity is $E_A(x, y)$. The point P where the electric field is to be determined is given by the spherical coordinates r_0, θ, φ. Since P is far away, the line drawn from dS to P may be taken as being parallel to OP and r as being nearly equal to r_0. We may replace r by r_0 in the denominator of eqn (5.261) and take it out of the integration sign, but for working out the phase the small differences between r and r_0 are significant. Thus for the exponent we shall use a better approximation:

$$\begin{aligned} r &= r_0 - \mathbf{R}\cdot\mathbf{i}_{r0} \\ &= r_0 - (x\mathbf{i}_x + y\mathbf{i}_y)\cdot(\sin\theta\cos\varphi\,\mathbf{i}_x + \sin\theta\sin\varphi\,\mathbf{i}_y + \cos\theta\,\mathbf{i}_z) \\ &= r_0 - (x\sin\theta\cos\varphi + y\sin\theta\sin\varphi). \end{aligned} \tag{5.262}$$

The electric field at point P may now be written as

$$E_\theta = -\mathrm{j}\frac{\beta_0}{2\pi}\frac{\exp(-\mathrm{j}\beta_0 r_0)}{r_0}\int_{-a/2}^{a/2}\int_{-b/2}^{b/2} E_A(x, y)\exp\{\mathrm{j}\beta_0(x\sin\theta\cos\varphi + y\sin\theta\sin\varphi)\}\,\mathrm{d}x\,\mathrm{d}y. \tag{5.263}$$

The above expression may be further simplified by assuming that

$$E_A(x, y) = f_1(x)f_2(y), \tag{5.264}$$

in which case

$$E_\theta = -\mathrm{j}\frac{\beta_0}{2\pi}\frac{\exp(-\mathrm{j}\beta_0 r_0)}{r_0}g_1(\theta, \varphi)g_2(\theta, \varphi), \tag{5.265}$$

where

$$g_1(\theta, \varphi) = \int_{-a/2}^{a/2} f_1(x)\exp(\mathrm{j}\beta_0 x\sin\theta\cos\varphi)\,\mathrm{d}x \tag{5.266}$$

and

$$g_2(\theta, \varphi) = \int_{-b/2}^{b/2} f_2(y) \exp(\mathrm{j}\beta_0 y \sin\theta \sin\varphi)\,\mathrm{d}y. \tag{5.267}$$

If we want to calculate the variation of E_θ in the x, z and y, z planes, we find that

$$E_\theta(\theta, 0) = -\mathrm{j}\frac{\beta_0}{2\pi}\frac{\exp(-\mathrm{j}\beta_0 r_0)}{r_0}\left(\int_{-b/2}^{b/2} f_2(y)\,\mathrm{d}y\right)\int_{-a/2}^{a/2} f_1(x)\exp(\mathrm{j}\beta_0 x \sin\theta)\,\mathrm{d}x \tag{5.268}$$

and

$$E_\theta(\theta, 90°) = -\mathrm{j}\frac{\beta_0}{2\pi}\frac{\exp(-\mathrm{j}\beta_0 r_0)}{r_0}\left(\int_{-a/2}^{a/2} f_1(x)\,\mathrm{d}x\right)\int_{-b/2}^{b/2} f_2(y)\exp(\mathrm{j}\beta_0 y \sin\theta)\,\mathrm{d}y, \tag{5.269}$$

i.e. $f_1(x)$ and $f_2(y)$ will determine the radiation pattern in the x, z and y, z planes respectively.

Let us restrict our attention to the x, z plane and discuss briefly the relationship between the field distribution (often called illumination function) $f_1(x)$, and the radiation pattern $g_1(\theta, 0)$. It may be seen from eqn (5.266) that this relationship takes the form of a Fourier transform. Hence we have at our disposal the whole mathematical apparatus of the Fourier transform for the analysis (what is the radiation pattern for a given field distribution) and synthesis (how to obtain the field distribution when the required radiation pattern is specified) of aperture antennas. I haven't said it in so many words so far, but it is fairly obvious that by establishing a field distribution over an aperture we have, in fact, got an antenna.

Can we work out the gain of such an antenna? Yes, we have got nearly everything we need. The power density in the direction of maximum radiation (θ_m, φ_m) may be obtained from eqn (5.263) as follows:

$$P_d(\theta_m, \varphi_m) = \frac{|E_\theta|^2}{2Z_0} = \frac{1}{2Z_0}\left(\frac{\beta_0}{2\pi r_0}\right)^2 |g_1(\theta_m, \varphi_m)|^2 |g_2(\theta_m, \varphi_m)|^2. \tag{5.270}$$

The total input power, i.e. the power crossing the aperture, may be worked out by assuming an input wave for which the laws of geometrical optics (particularly eqn (5.250)) are valid, yielding

$$P = \frac{1}{2}\int \frac{|E_A|^2}{Z_0}\mathbf{i}_p \cdot \mathrm{d}\mathbf{S} = \frac{\mathbf{i}_p \cdot \mathbf{i}_n}{2Z_0}\int |E_A|^2\,\mathrm{d}S, \tag{5.271}$$

where $\mathbf{i}_n$ is the unit vector normal to the aperture. If we feed this power into an isotropic antenna the power density at a distance r_0 is

$$P_{d,is} = \frac{\mathbf{i}_p \cdot \mathbf{i}_n}{2Z_0}\frac{1}{4\pi r_0^2}\int |E_A|^2\,\mathrm{d}S. \tag{5.272}$$

By definition, the gain is given by

$$G = \frac{P_d(\theta_m, \varphi_m)}{P_{d,is}} = \frac{\beta_0^2}{\pi \mathbf{i}_p \cdot \mathbf{i}_n} \frac{|g_1(\theta_m, \varphi_m)|^2 |g_2(\theta_m, \varphi_m)|^2}{\int |E_A|^2 \, dS}. \tag{5.273}$$

This is a general formula for gain valid for any field distribution. In the special case when E_A is produced by a plane wave incident perpendicularly upon the aperture

$$E_A = E_0 = \text{constant}, \qquad \mathbf{i}_p \cdot \mathbf{i}_n = 1. \tag{5.274}$$

What will be the direction of maximum radiation? If the aperture is an equiphase surface then in the $\theta = 0$ direction the radiation from each surface element adds algebraically. In any other direction some phase factors are involved, reducing the total sum, hence $\theta_m = 0$. Substituting the above results into eqn (5.273) and using eqns (5.266) and (5.267) we get for the gain of a uniformly illuminated aperture

$$G = \frac{\beta_0^2}{\pi} \frac{E_0^2 a^2 b^2}{E_0^2 ab} = \frac{4\pi}{\lambda^2} ab = \frac{4\pi A}{\lambda^2}, \tag{5.275}$$

where A is the area of the aperture.

We got, at least, a simple formula. The gain is equal to 4π times the area of the aperture measured in wavelengths. If you want high gain (that is a highly directional antenna) you need a large aperture. How can one realize an aperture antenna? The usual realization is in the form of a reflector fed by a primary radiator. We did in fact carry out a little design of our own in the previous section where we derived the shape of a two-dimensional reflector required to produce a plane wavefront. In a practical case it would take the form of a cylindrical parabola of finite length illuminated by a line-source feed as shown in Fig. 5.35. The field distribution is constant in the x direction and tapered in the y direction (c.f. curve (b) of Fig. 5.32). In any case if the radiation pattern of the line source is given both $f_1(x)$ and $f_2(y)$ are calculable with the aid of geometrical optics, and then eqns (5.266) and (5.267) yield

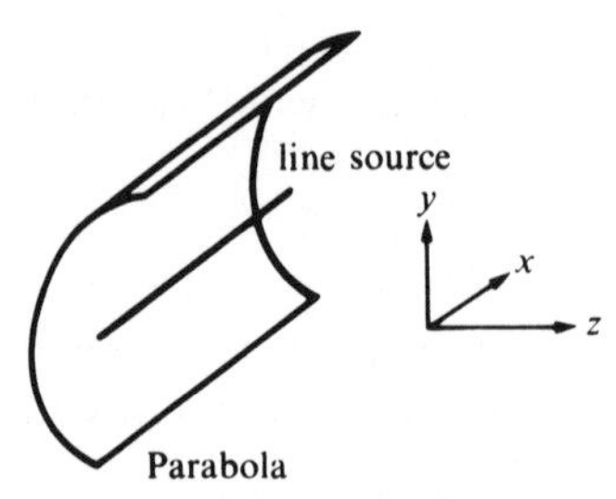

FIG. 5.35. A parabolic reflector as an example of realizing a field distribution over a rectangular aperture.

the radiation pattern of the reflector, whereas eqn (5.273) will provide the gain.

I have discussed Huygens' principle and some aspects of the design of aperture antennas in order to show you both the simplicity and the versatility of electromagnetic theory. We find an approximation that fits the geometry. Both geometrical optics and Huygens' principle are consequences of Maxwell's equations but they lead to different physical pictures and to radically different formulae. Note the division of labour between them. The reflector surface may be designed with the aid of geometrical optics because it is large in comparison with the wavelength, and the far-field pattern of the reflector may be calculated from a simplified version of Huygens' principle (all spherical waves parallel) because the observation point is far away.

5.20. Diffraction by a thin half-infinite plane

In this section we shall be concerned with the solution of an abstract boundary-value problem. We shall postulate a semi-infinite metal plate that is both infinitely thin and possesses infinite conductivity. Such plates of course don't exist in real life, but as long as the plate thickness is small in comparison with the wavelength, the above abstraction may be regarded as valid.

Let a plane wave be incident upon the plate at an angle α as shown in Fig. 5.36(a). If we assume in addition that the wave is so polarized that the electric field is parallel with the z axis, then the problem is two-dimensional, characterized by the polar coordinates R, φ. What are the conditions to be satisfied? First of all Maxwell's equations, and then the boundary condition that E_z must vanish on both faces of the plate. There are some other conditions too, concerned with the physical reality of the solution, e.g. a

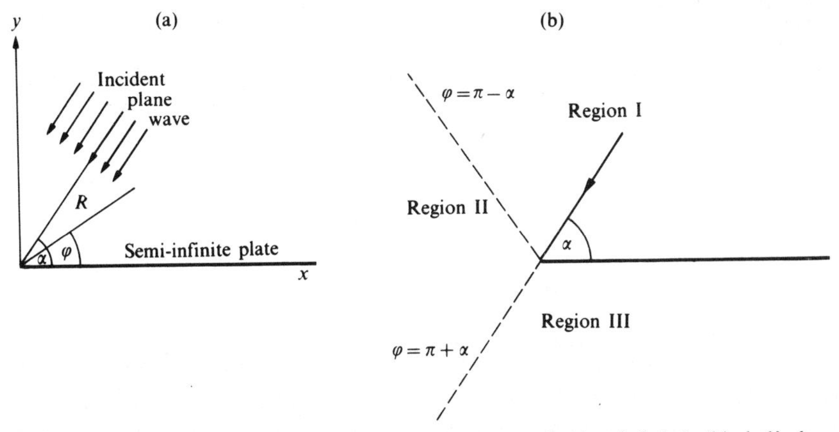

FIG. 5.36. (a) A plane wave incident upon a perfecting conducting, infinitely thin half-plane. (b) Regions classified according to geometrical optics.

solution that satisfied both the differential equations and the boundary conditions but gave an electric field that increased indefinitely towards infinity would obviously be of no use. Anyway, it is not my intention to show how to eliminate unwanted solutions nor will I indicate the main steps in the derivation. It is the old story again. The derivation is too difficult, and unless you have some absorbing interest in complex analysis it is not particularly entertaining. Nevertheless, it is of great interest that such a problem can be solved in closed form, and its further claim to fame is that (as you will see in the next section) it triggered off a line of thought leading to an entirely new physical picture of the diffraction phenomenon.

Using the polar coordinates R, φ the incident wave propagating in the $(\alpha+\pi)$ direction may be described as follows:

$$\begin{aligned} E_{z0} &= E_0 \exp(-\mathrm{j}\beta_0 R_0) \\ &= E_0 \exp[-\mathrm{j}\beta_0\{\cos(\alpha+\pi)\mathbf{i}_x + \sin(\alpha+\pi)\mathbf{i}_y\}R(\cos\varphi\mathbf{i}_x + \sin\varphi\mathbf{i}_y)] \\ &= E_0 \exp\{\mathrm{j}\beta_0 R\cos(\varphi-\alpha)\}. \end{aligned} \tag{5.276}$$

The electric field modified by the presence of the plate must obey the two-dimensional wave equation:

$$\frac{\partial^2 E_z}{\partial R^2} + \frac{1}{R}\frac{\partial E_z}{\partial R} + \frac{1}{R^2}\frac{\partial^2 E_z}{\partial \varphi^2} + \beta_0^2 E_z = 0 \tag{5.277}$$

and the boundary conditions

$$E_z = 0 \quad \text{for} \quad \varphi = 0 \quad \text{and} \quad \varphi = 2\pi. \tag{5.278}$$

The solution is given by the formula

$$E_z = E_0\{V(R,\varphi-\alpha) - V(R,\varphi+\alpha)\}, \tag{5.279}$$

where

$$V(R,\psi) = \frac{1+\mathrm{j}}{2}\exp(\mathrm{j}\beta_0 R\cos\psi)F\{h(R,\psi)\} \tag{5.280}$$

and

$$\begin{aligned} F\{h(R,\psi)\} &= \int_{-\infty}^{h(R,\psi)} \exp\left(-\mathrm{j}\frac{\pi}{2}u^2\right)\mathrm{d}u, \\ h(R,\psi) &= 2\left(\frac{\beta_0 R}{\pi}\right)^{\frac{1}{2}}\cos\frac{\psi}{2}. \end{aligned} \tag{5.281}$$

If you have any doubts prove first (Example 5.29) that eqns (5.279) to (5.281) satisfy eqn (5.277). The proof is straightforward, though a little lengthy. Does our solution satisfy the boundary conditions? Yes, it does. Is another lengthy proof required? No, not at all. Since the angular dependence of V is due to the trigonometric functions $\cos \psi$ and $\cos \psi/2$ and since

$$\cos(\varphi-\alpha)=\cos(\varphi+\alpha) \quad \text{and} \quad \cos\frac{\varphi-\alpha}{2}=\cos\frac{\varphi+\alpha}{2} \tag{5.282}$$

for $\varphi=0$ and 2π, it follows that on the boundaries

$$V(R, \varphi-\alpha)=V(R, \varphi+\alpha), \tag{5.283}$$

and E_z vanishes.

There are a number of further conclusions to be drawn but, before I go further, I have to say a few words about the properties of the functions given in eqn (5.281). If you look up the *Tables of higher functions* (see footnote on p. 158) you will find the so called Fresnel integrals defined as

$$C\left(\frac{\pi}{2}z^2\right)=\int_0^z \cos\frac{\pi}{2}u^2\,du \quad \text{and} \quad S\left(\frac{\pi}{2}z^2\right)=\int_0^z \sin\frac{\pi}{2}u^2\,du. \tag{5.284}$$

They are tabulated there, and asymptotic formulae for large values of the argument are also given. Since both of the above functions are symmetric in z and their limiting values are

$$C(\infty)=S(\infty)=\tfrac{1}{2}, \tag{5.285}$$

the function $F(h)$ may be written as

$$\begin{aligned} F(h)&=\int_{-\infty}^{h} \exp\left(-j\frac{\pi}{2}u^2\right) du=\int_{-\infty}^{0} \exp\left(-j\frac{\pi}{2}u^2\right) du+\int_0^h \exp\left(-j\frac{\pi}{2}u^2\right) du \\ &=\frac{1-j}{2}+C\left(\frac{\pi}{2}h^2\right)-jS\left(\frac{\pi}{2}h^2\right). \end{aligned} \tag{5.286}$$

Thus if you want the value of the electric field at any point in space you can work it out with the aid of the Tables given.

The asymptotic formula valid for $|h|\gg 1$ is of the form

$$\int_0^h \exp\left(-j\frac{\pi}{2}u^2\right) du \simeq \frac{1-j}{2}+j\frac{\exp(-j(\pi/2)h^2)}{\pi h}. \tag{5.287}$$

We are ready now to discuss the solution of our boundary-value problem. A convenient starting point is to divide the space into three regions on the basis of geometrical optics, as shown in Fig. 5.36(b).

$$\left.\begin{array}{ll} 0<\varphi<\pi-\alpha & \text{region I,} \\ \pi-\alpha<\varphi<\pi+\alpha & \text{region II,} \\ \pi+\alpha<\varphi<2\pi & \text{region III.} \end{array}\right\} \tag{5.288}$$

There are two waves in region I (incident plus reflected), one wave in region II (incident), and no wave at all in region III. If our solution is correct and if geometrical optics is indeed a $\lambda \to 0$ limit of electromagnetic theory then the existence of these three regions should follow from eqns (5.279) to (5.281).

Let us look at region III first, where

$$\frac{\pi}{2} < \frac{\varphi - \alpha}{2} < \pi - \frac{\alpha}{2} \quad \text{and} \quad \frac{\pi}{2} + \alpha < \frac{\varphi + \alpha}{2} < \pi + \frac{\alpha}{2}. \tag{5.289}$$

Provided that $\alpha < \pi$ (and of course the angle of incidence cannot exceed π because that would lead to the same problem upside down) it follows that both $\cos(\varphi - \alpha)/2$ and $\cos(\varphi + \alpha)/2$ are negative, hence

$$h(R, \varphi - \alpha) < 0 \quad \text{and} \quad h(R, \varphi + \alpha) < 0 \tag{5.290}$$

and

$$\lim_{\beta_0 \to \infty} h(R, \varphi - \alpha) \to -\infty \quad \text{and} \quad \lim_{\beta_0 \to \infty} h(R, \varphi + \alpha) \to -\infty. \tag{5.291}$$

But

$$\lim_{h \to -\infty} F(h) = 0, \tag{5.292}$$

thus

$$E_z = E_0(0 - 0) = 0. \tag{5.293}$$

It has now been shown that in the limit of $\lambda \to 0$ the electric field vanishes in region III.

In region II

$$\frac{\pi}{2} - \alpha < \frac{\varphi - \alpha}{2} < \frac{\pi}{2} \quad \text{and} \quad \frac{\pi}{2} < \frac{\varphi + \alpha}{2} < \frac{\pi}{2} + \alpha, \tag{5.294}$$

hence

$$\cos\frac{\varphi - \alpha}{2} > 0 \quad \text{and} \quad \cos\frac{\varphi + \alpha}{2} < 0 \tag{5.295}$$

and

$$\lim_{\beta_0 \to \infty} h(R, \varphi - \alpha) \to \infty \quad \text{and} \quad \lim_{\beta_0 \to \infty} h(R, \varphi + \alpha) \to -\infty. \tag{5.296}$$

In view of eqn (5.292) and considering that

$$\lim_{h \to \infty} F(h) = \frac{1 - \mathrm{j}}{2}, \tag{5.297}$$

we get

$$E_z = E_0\{V(R, \varphi-\alpha)-0\} = E_0\frac{1+\mathrm{j}}{2}\exp\{\mathrm{j}\beta_0 R\cos(\varphi-\alpha)\}\frac{1-\mathrm{j}}{2}$$

$$= E_0\exp\{\mathrm{j}\beta_0 R\cos(\varphi-\alpha)\}, \tag{5.298}$$

i.e. the field in region II is the same as the incident field.

One can similarly show the presence of the incident and reflected waves in region I.

We shall now consider the case when the wavelength is finite but $\beta_0 R \gg 1$. In region III h is negative, hence

$$F(h) = \int_{-\infty}^{0}\exp\left(-\mathrm{j}\frac{\pi}{2}u^2\right)\mathrm{d}u - \int_{-|h|}^{0}\exp\left(-\mathrm{j}\frac{\pi}{2}u^2\right)\mathrm{d}u$$

$$= \frac{1-\mathrm{j}}{2} - \int_{0}^{|h|}\exp\left(-\mathrm{j}\frac{\pi}{2}u^2\right)\mathrm{d}u. \tag{5.299}$$

If in addition $|h| \gg 1$ we may employ eqn (5.287), yielding

$$F(h) = -\frac{\mathrm{j}}{\pi|h|}\exp\left(-\mathrm{j}\frac{\pi}{2}h^2\right), \tag{5.300}$$

leading to

$$V(R, \psi) = \frac{1+\mathrm{j}}{2}\left(-\frac{\mathrm{j}}{\pi|h|}\right)\exp\{\mathrm{j}(\beta_0 R\cos\psi - 2\beta_0 R\cos^2\psi/2)\}$$

$$= \frac{1-\mathrm{j}}{2\pi|h|}\exp(-\mathrm{j}\beta_0 R). \tag{5.301}$$

Substituting eqn (5.301) into (5.279) we get for the asymptotic solution in region III

$$E_z = E_0\frac{\mathrm{j}-1}{4(\pi\beta_0 R)^{\frac{1}{2}}}\exp(-\mathrm{j}\beta_0 R)\left\{\frac{1}{\cos[(\varphi-\alpha)/2]} - \frac{1}{\cos[(\varphi+\alpha)/2]}\right\}. \tag{5.302}$$

This is a cylindrical wave. We have so far talked about plane waves and spherical waves but not about cylindrical waves. We shall rectify that omission now.

A cylindrical wave spreads out radially, hence it should possess a phase factor $\exp(-\mathrm{j}\beta_0 R)$. If we assume that power moves also radially outwards then power density must decay as $1/R$. This may be proved by the following argument. If the power radiated out per unit length in an angle $\mathrm{d}\varphi$ is $P\,\mathrm{d}\varphi$ then at a distance R it will cross an arc $R\,\mathrm{d}\varphi$, yielding for the power density

$$P_\mathrm{d} = \frac{P\,\mathrm{d}\varphi}{R\,\mathrm{d}\varphi} = \frac{P}{R}. \tag{5.303}$$

Assuming further that the power flow is proportional to the square of the electric field we can immediately state that E_z must vary with R as

$$E_z \sim R^{-\frac{1}{2}} \exp(-j\beta_0 R), \tag{5.304}$$

in agreement with eqn (5.294).

The angular dependence of E_z is given by the braces in eqn (5.301). One would expect that the 'deeper' we move into the shadow region, that is the nearer φ is to 2π, the smaller will be the amplitude of the electric field. This is indeed the case, as may be seen in Fig. 5.37, where the angular dependence is plotted for an incident angle, $\alpha = 60°$.

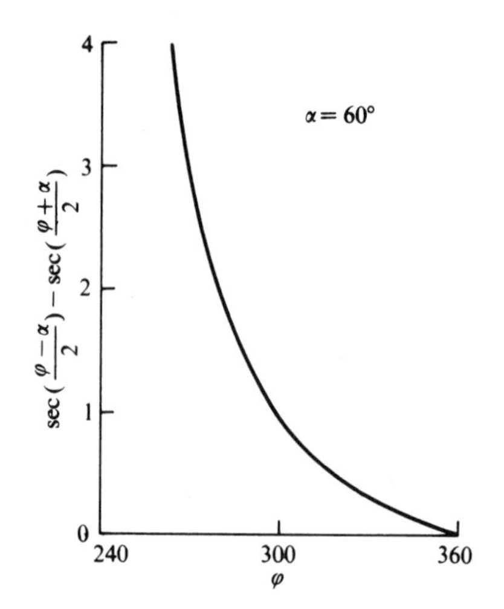

FIG. 5.37. The variation of the electric field with angle in the shadow region.

Note that

$$\frac{1}{\cos[(\varphi-\alpha)/2]} \to \infty \quad \text{when} \quad \varphi \to \alpha + \pi, \tag{5.305}$$

i.e. E_z tends to infinity at the boundary of the shadow region. Is this possible? No, E_z must be finite at the shadow boundary. Where is the mistake then? The mistake was committed by using eqn (5.301) in a region where it is not valid. In the vicinity of $\varphi = \alpha + \pi$, $\cos[(\varphi-\alpha)/2] \cong 0$ and the condition $\beta_0 R \gg 1$ is no longer sufficient for regarding $|h| \gg 1$. Hence for these values of φ the asymptotic expansion cannot be used, and one needs to return to the exact formula.

Finally, we shall work out the field in region II for $\beta_0 R \gg 1$. Then $h(R, \varphi-\alpha) > 0$ and $h(R, \varphi+\alpha) > 0$, which follows from eqn (5.295). Accordingly, we can use our previously obtained result (eqn (5.301)) for

$h(R, \varphi+\alpha)$ but for $h(R, \varphi-\alpha)$ we need a new derivation. Since

$$F(h)=\frac{1-\mathrm{j}}{2}+\int_0^h \exp\left(-\mathrm{j}\frac{\pi}{2}u^2\right)\mathrm{d}u, \tag{5.306}$$

we get for $h \gg 1$ with the aid of eqn (5.279),

$$F(h)=1-\mathrm{j}+\frac{\mathrm{j}}{\pi h}\exp\left(-\mathrm{j}\frac{\pi}{2}h^2\right). \tag{5.307}$$

Substituting now both for $V(R, \varphi-\alpha)$ and for $V(R, \varphi+\alpha)$ into eqn (5.279) we obtain after a little rearrangement

$$E_z=E_{z0}+E_0\frac{\mathrm{j}-1}{4(\beta_0\pi R)^{\frac{1}{2}}}\exp(-\mathrm{j}\beta_0 R)\left\{\frac{1}{\cos[(\varphi-\alpha)/2]}-\frac{1}{\cos[(\varphi+\alpha)/2]}\right\} \tag{5.308}$$

As may be seen there is a 'diffraction field' to be added to the incident field. We shall further investigate some aspects of the asymptotic solutions in the next section. If you wish to find out how to arrive at eqn (5.279) or would like to see the solution of some other interesting diffraction problems, look up Sommerfeld's *Optics*.†

5.21. The geometrical theory of diffraction

We shall start the discussion by re-casting some of the results obtained in the previous section in more convenient forms. We shall introduce the notation

$$E_\mathrm{d}=E_0\frac{\mathrm{j}-1}{4(\beta_0\pi R)^{\frac{1}{2}}}\exp(-\mathrm{j}\beta_0 R)\left\{\frac{1}{\cos[(\varphi-\alpha)/2]}-\frac{1}{\cos[(\varphi+\alpha)/2]}\right\}, \tag{5.309}$$

with which eqns (5.301) and (5.308) may be rewritten as

$$\begin{aligned} E_z&=0+E_\mathrm{d} \qquad \pi+\alpha<\varphi<2\pi \\ E_z&=E_{z0}+E_\mathrm{d} \qquad \pi-\alpha<\varphi<\pi+\alpha. \end{aligned} \tag{5.310}$$

Thus both in region III and in region II (Fig. 5.36(b)) the electric field at a sufficiently large distance may be represented as the sum of the geometrical optics field (zero in region III and E_{z0} in region II) and of another field (we shall call it the diffracted field) due to the presence of the plate. Note further that this diffracted field is of the form of a cylindrical wave whose axis coincides with the edge of the plate. For emphasizing further this fact let us introduce

$$D(\varphi, \alpha)=\frac{\mathrm{j}-1}{4(\beta_0\pi)^{\frac{1}{2}}}\left\{\frac{1}{\cos[(\varphi-\alpha)/2]}-\frac{1}{\cos[(\varphi+\alpha)/2]}\right\}, \tag{5.311}$$

† A. SOMMERFELD (1964). *Optics*. Academic Press, New York.

so that the diffracted field may be written in the form

$$E_d = E_0 D(\varphi, \alpha) \frac{\exp(-j\beta_0 R)}{R^{\frac{1}{2}}}. \tag{5.312}$$

We may now interpret the above equation as follows. E_0 is the field incident upon the edge of the plate located at $R=0$. This incident ray gives rise to diffracted rays moving radially outwards in *all* directions. Power is however *not* uniformly distributed. The power density in any given direction depends on the factor $|D(\varphi, \alpha)|^2$, where $D(\varphi, \alpha)$ is the *diffraction coefficient*. What is the advantage of the above description? Simplicity. Instead of looking for the solution of Maxwell's equations subject to the boundary conditions we just consider the geometrical optics field and add to it the field due to the rays produced by the edge of the plate. 'Very simple indeed', you might say, 'but what about the diffraction coefficient? In order to obtain the formula for $D(\varphi, \alpha)$ we need to solve the boundary-value problem anyway. So the introduction of a diffraction coefficient might be an elegant way of presenting the results but it does not reduce the amount of mathematical work required.'

In a rigorous world the above argument would be valid. The description in terms of diffracted rays would have a mere aesthetic value. In the real world however the value of simplicity is not only in the eye of the beholder. A simple physical picture usually leads to valid approximations for solving more complex problems. Let us take for example the problem shown in Fig. 5.38(a). We have now two infinitely thin, infinitely conducting plates parallel to each other, and we assume again a plane wave incident at an angle α_1. How can we work out the electric field at point P? Solve Maxwell's equations subject to the boundary conditions. Yes, but how? No one has so far been able to express the solution of the above problem in closed form. One could of course turn to numerical techniques, but that is tiresome and

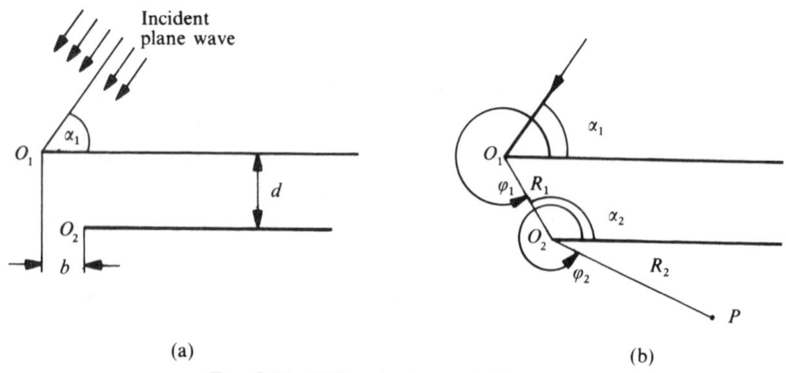

FIG. 5.38. Diffraction by two half-planes.

expensive. So what shall we do? Use our new theory. The incident ray at O_1 excites diffracted rays with amplitudes $E_0D(\alpha_1, \varphi_1)$. The diffracted rays diverge hence the amplitude of the ray arriving at O_2 (Fig. 5.38(b)) is

$$E_{z2} = E_0D(\alpha_1, \varphi_1)\frac{\exp(-j\beta_0R_1)}{R_1^{\frac{1}{2}}}. \tag{5.313}$$

This diffracted ray will give rise to another family of diffracted rays when hitting the edge of the second plate at O_2. The amplitude of the ray travelling towards point P is $E_{z2}D(\alpha_2, \varphi_2)$, hence the electric field at P is

$$\begin{aligned} E_{z3} &= E_{z2}D(\alpha_2, \varphi_2)\frac{\exp(-j\beta_0R_2)}{R_2^{\frac{1}{2}}} \\ &= E_0D(\alpha_1, \varphi_1)D(\alpha_2, \varphi_2)\frac{\exp\{-j\beta_0(R_1+R_2)\}}{(R_1R_2)^{\frac{1}{2}}}. \end{aligned} \tag{5.314}$$

You couldn't have a simpler solution for a diffraction problem.

Let us next take the classical problem Thomas Young was investigating (see quotation at the beginning of the chapter). Spherical waves emitted by a source at A are incident upon an aperture (Fig. 5.39(a)). How can we find

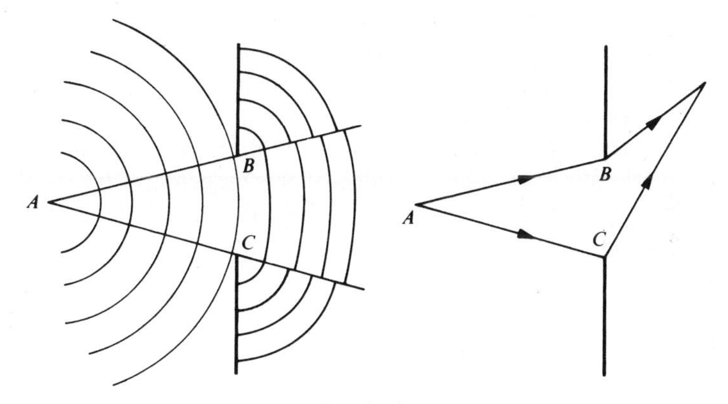

FIG. 5.39. Diffraction by an aperture.

the field in the shadow region? 'Faint radiations on each side will diverge from B and C as centres', suggested Young. These are exactly our diffracted rays. In our formulation we say that for finding the field intensity in the shadow region we need to add the contributions (in the right phase) from two diffracted rays as shown in Fig. 5.39(b). There is of course nothing special in infinitely thin plates. The edge of a wedge should also give rise to diffracted rays (Fig. 5.40(a)). That would necessitate the solution of another boundary-value problem, but having done that and having derived the wedge diffraction coefficients we could now solve problems like that shown in Fig. 5.40(b). There are a number of other kind of discontinuities giving rise to diffracted

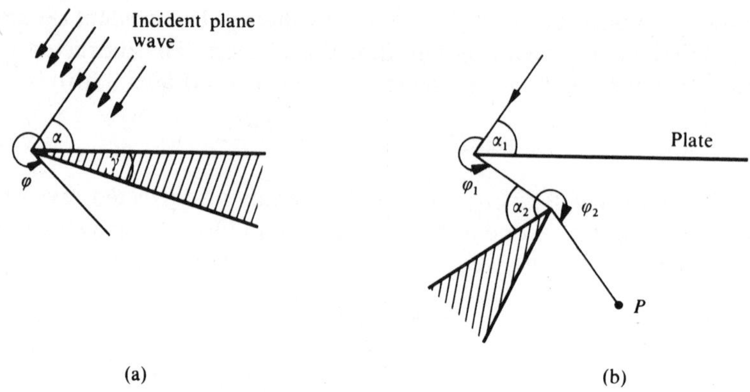

FIG. 5.40. Diffraction by wedges.

rays. I do not wish to enumerate them here; I'd rather give you a practical example concerned with the back-radiation of a parabolic reflector.

As we discussed before, the radiation pattern of an aperture antenna may be calculated from eqn (5.263). Note however that the whole procedure outlined in Section 5.18 was based on the assumption that contributions from each surface element of the aperture may be added. Unfortunately, this construction is restricted to determining the field radiated forward. Take for example the parabolic cylinder of Fig. 5.35. Backward radiation from the aperture is blocked by the reflector hence the field in the reverse direction cannot be determined on the basis of Huygens' principle. Our ray theory will however work. We may take the edge of the reflector as the edge of an infinite plate.† Thus for a direction $\theta > \pi/2$ we add the rays diffracted by the upper and lower edges as shown in Fig. 5.41.

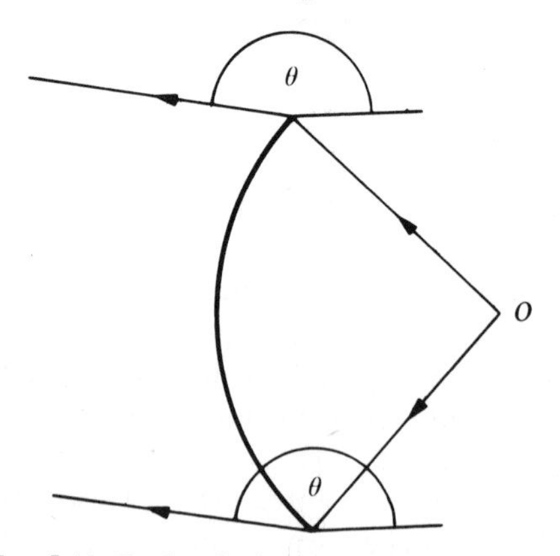

FIG. 5.41. Back radiation by the edge of a reflector.

† That's the beauty of the whole thing; we are concerned only with *local* phenomena. An incident ray is not too inquisitive, it explores only the immediate vicinity of the point of diffraction. If that looks like the edge of an infinite plate the ray is happy to diffract as if it *was* an infinite plate.

If our parabolic cylinder was infinitely large the field would spread as $1/R^{\frac{1}{2}}$. Since it is of finite axial dimension the $1/R^{\frac{1}{2}}$ spread is confined to the vicinity of the reflector. Further away the wave must look as a spherical one decaying as $1/R$. Since all antenna structures are finite (in contrast to the two-dimensional cases discussed for simplicity) we need to incorporate into our equations this more general type of R dependence. I am not going to derive that here (not because it's difficult but it's a question of detail) if you are interested look up J. B. Keller (1957).†

My aim in this short section was to give you an illustration of the origin and power of this new type of diffraction theory. It lay dormant for a century and a half; it owes its resurrection to Keller (his basic paper is cited above) who called it the geometrical theory of diffraction. It is one of the 'growth' subjects in electromagnetic theory today appealing both to the theoretician and to the engineer.

5.22. Supergain and superresolution

As you may have noticed, 'super' has become a fashionable prefix. It has probably started with Nietzsche who coined the word *Übermensch*, and then came Shaw with his *Man and superman*, and Freud with his 'superego'. I greatly deplore the use of this prefix by 'arts' people. They have at their disposal the full vocabulary of the English language. They should always be able to find a word to fit any occasion. A scientist, however, may easily find himself in an unpleasant predicament. Although he is free to invent new words (like enthropy or conductivity) or give new meanings to old words (like gain or resolution), once he introduces a definition or calculates the limiting value of something, he is stuck.

I shall tell you as an example the story of gain and supergain. As you will remember we defined the gain of an antenna in Section 5.16 and derived an expression for the gain of an aperture antenna in Section 5.18. For uniform field distribution in the aperture we found that the gain came to $4\pi A/\lambda^2$. Had we had more time I would have shown, on the basis of eqn (5.273), that $4\pi A/\lambda^2$ is the maximum gain an aperture antenna may have for *any* field distribution. Very good, so we have a definition for gain and an expression for the maximum gain.

All derivations however, are generously sprinkled with assumptions, and the results are only as good as the assumptions. If we drop the assumption that the wave crossing the aperture is a local plane wave then the maximum gain for an aperture of given size turns out to be infinitely large. I shall talk presently about the engineering problems; they are quite difficult, but you should not underestimate the semantic problem either. What do you call the gain that is higher than the maximum? You introduce the word 'supergain', and everything is saved. The old maximum is still a maximum valid under

† J. B. KELLER (1957). Diffraction by an aperture. *J. appl. Phys.* **28** 426–44.

certain conditions, but any gain higher than that will be called supergain.

How can one prove that the antenna gain has no upper limit? It is sufficient to investigate the radiation pattern in one plane. Substituting $\varphi = 0$ into eqn (5.266) we get for the radiation pattern in the x, z plane

$$g_1(\theta, 0) = \int_{-a/2}^{a/2} f_1(x)\, e^{jux}\, dx, \qquad u = \beta_0 \sin\theta, \tag{5.315}$$

i.e. the radiation pattern is the Fourier transform of the field distribution along the x direction.

It follows from the properties of the Fourier transform (the mathematical proof is difficult, requiring a deep plunge into function theory†) that any specified $g_1(\theta, 0)$ may be realized with arbitrary accuracy by using an aperture antenna of finite size. A narrow beam means high gain, hence an arbitrarily narrow beam means that arbitrarily high gain may be achieved.

The implications are far-reaching, I might say bewildering. Are we wasting millions of pounds by building enormous antennas for radio astronomy and satellite communications? Couldn't we use little pocket antennas instead? Part of the answer comes from the self-same theory. Any gain higher than the normal one must be paid for by increased tolerance sensitivity (the field distribution in the aperture must be kept within strictly prescribed limits) and reduced bandwidth.

There have been several successful attempts at producing supergain antennas, but the gain achieved was not much above the normal, whereas the design effort was quite substantial. Expert opinion tends toward the view that supergain antennas are not practicable, and that is probably the right verdict. Nevertheless, this is still a controversial field. One day, one man might bring out a new antenna construction which might be still far from a pocket antenna but could reduce the required diameter from 50 m to (say) 5 m. And I'm sure you will agree that it is an economy worth having.

Going over to a somewhat different subject, no doubt familiar to you, I shall give below the formula for the limiting resolution of an optical microscope:

$$\text{limiting resolution} = \frac{\lambda}{2n \sin\alpha}, \tag{5.316}$$

where 2α is the angle of opening of the object ray cone and $n \sin\alpha$ is known as the numerical aperture. Its value rarely exceeds 1·6, so the resolution is of the order of the wavelength. The above formula (derived by Abbe) rests on the diffracting properties of light, in other words it is based on the properties of radiating electromagnetic waves.

† See, for example, R. Kovacs and L. Solymar (1956). Theory of aperture aerials based on the properties of entire functions of the exponential type. *Acta phys. hung.* **6**, 161–84.

If you achieve a resolution better than the optimum, and if you adopt the semantic strategy developed in this section, then you will refer to your achievement as *superresolution.* Anyway this is the word coined by Ash whose basic idea (in 1972) contained the following three ingredients:

1. illuminate the object by a non-radiative field;
2. move the object so that a small part of it may be explored at a time;
3. vibrate the object so that the useful information may be more efficiently extracted.

Let us now go into more details. The position of the object relative to the illuminating electromagnetic field is shown schematically in Fig. 5.42. A small hole of diameter $2r_0 \ll \lambda$ is cut out of the wall of an electromagnetic resonator. Since the hole is small in comparison with the wavelength it is in the same category as the elementary dipole discussed in Section 5.16. In the immediate vicinity of the hole the fields are of the static type, which decay with the third power of distance. Hence the fields at the object and the energy stored in the intervening space depend on the shape of the object. Obviously, in position (a) there is less interference with the fields than in position (b) (see Figs. 5.42(a) and (b)). Thus by moving the object in front of

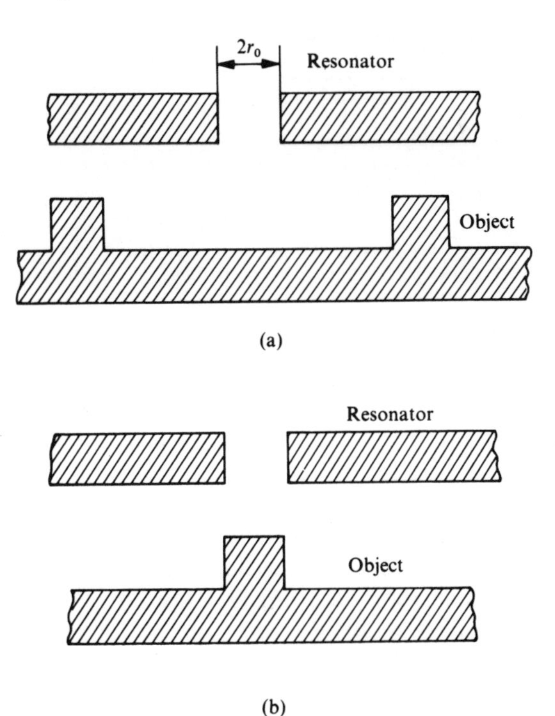

FIG. 5.42. The object 'seen' in two different positions.

the hole we may obtain information about that point of the object nearest to the hole.

Next problem is how to extract the information available. In practice the simplest thing proved to be to measure the signal reflected by the resonator. This appears to be a sufficiently sensitive measure provided we choose the frequency f_0 of the signal to correspond to a steep part of the resonant curve.

A schematic diagram of the basic arrangement used at a wavelength of about 3 cm is shown in Fig. 5.43. The illumination of the object is achieved

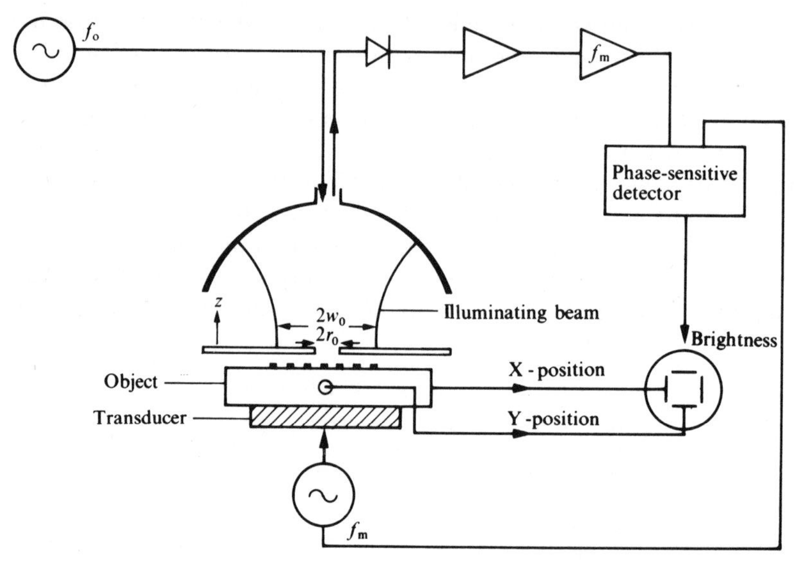

FIG. 5.43. Schematic representation of the operation of the scanning microscope. (From E. A. Ash and G. Nicholls (1972). *Nature, Lond.* (June 30) 237.)

through the hole in the open resonator. The information, as mentioned before, is contained in the reflected signal. Note however that small changes in reflection may occur for a number of other reasons to which we may refer as background noise. In order to enhance the useful information the object is mounted on a transducer that is vibrated at an ultrasonic frequency. Hence the reflected signal will have a component varying at f_m, the frequency of the ultrasound. The detection system shown in Fig. 5.43 extracts this component and displays the information on an oscilloscope (the larger the f_m component, the brighter the screen), where the image of the object will duly appear. Ash and Nicholls have reported resolution as small as $\lambda/60$ and there is no reason to assume that they were anywhere near the limit.†

† E. A. ASH and G. NICHOLLS (1972) *Nature, Lond.* **237** 510–12.

Gain and resolution are not the only two cases when well-established limits crumbled under the assault of electrical engineers. Another interesting story is that of the noise figure of a microwave tube. Engineering ingenuity found means of reducing the noise produced by a hot cathode to an incredibly low value, baffling thereby all theoreticians engaged in the art of thermodynamics. I might tell that story in another lecture course. I shan't give any more examples now, I think I've made my point. Don't be overwhelmed by existing proofs showing that such-and-such a limit cannot be transgressed. Look always at the assumptions.

Examples 5

5.1. A straight metal wire of conductivity σ and circular cross-sectional area S_0 carries a steady current I. Determine the direction and magnitude of the Poynting vector at the surface of the wire. Integrate the normal component of the Poynting vector over the surface of the wire for a segment of length l and compare your result with the resistive heating produced in this segment.

5.2. Describe qualitatively the process of charging up a capacitor in terms of the Poynting vector.

5.3. The electric field in a plane wave propagating in the z direction is as follows

$$\mathbf{E} = E_1 \cos(\omega t - \beta_0 z)\mathbf{i}_x + E_2 \sin(\omega t - \beta_0 z)\mathbf{i}_y \tag{5.317}$$

where ω is the angular frequency, β_0 is the free space propagation coefficient, and E_1 and E_2 are constants.

(i) Find the corresponding magnetic field,

(ii) Show that for a fixed value of z the end of the electric field vector moves on an ellipse as a function of time.

5.4. What is the skin depth at 50 Hz? Is it relevant for power transmission? Would it effect the performance of iron cored transformers?

5.5. Take a plane wave that propagates in the z direction and has only E_x and H_y components. When the wave enters a good conductor the electric field decays as given by eqn (5.107).

(i) Show that the wave impedance in the conductor is of the form

$$Z_c = \frac{E_x}{H_y} = \left(\mathrm{j}\frac{\omega\mu}{\sigma}\right)^{\frac{1}{2}}. \tag{5.318}$$

(ii) Derive the Poynting vector.

5.6. Take again a plane wave propagating in a good conductor as in Example 5.5. In view of the discussion in Section 5.12 the force on a $dx\,dy$ element of the material is

$$F = \tfrac{1}{2}(\rho\, dx\, dy\, dz) v_x \mu_0 H_y^* \tag{5.319}$$

where ρ is the charge density. Neglecting displacement current and noting that ρv_x is equal to the current density in the x direction show that the radiation pressure acting upon a semi-infinite $(0 < z < \infty)$ conductor is equal to

$$p = \tfrac{1}{4}\mu_0 |H_y(0)|^2 \tag{5.320}$$

where $H_y(0)$ is the magnetic field in the conductor at $z = 0$.

5.7. A transverse electromagnetic (TEM) wave propagates in the z direction with a propagation coefficient β in a lossless material characterized by the constants ε and μ. Show that $\beta = \omega(\varepsilon\mu)^{\frac{1}{2}}$ and that all the field components satisfy Laplace's equation.

5.8. A TE_{mn} wave propagating in the z direction in a hollow waveguide lacks an E_z component but, in the general case, all the other field components (E_x, E_y, H_x, H_y, H_z) are finite. Assume a propagation coefficient β and show that eqns (5.108) and (5.109) lead to the differential equation

$$\frac{\partial^2 H_z}{\partial x^2}+\frac{\partial^2 H_z}{\partial y^2}+(\beta_0^2-\beta^2)H_z=0. \tag{5.321}$$

Express H_x, H_y, E_x, and E_y with the aid of H_z.

5.9. (i) Assume that H_z in Example 5.8 may be expressed as

$$H_z = X(x)Y(y) \tag{5.322}$$

and show that eqn (5.321) will reduce to the differential equations

$$\frac{\mathrm{d}^2X}{\mathrm{d}x^2}+\beta_x^2X=0 \quad \text{and} \quad \frac{\mathrm{d}^2Y}{\mathrm{d}x^2}+\beta_y^2Y=0, \tag{5.323}$$

where

$$\beta_0^2-\beta_x^2-\beta_y^2=\beta^2. \tag{5.324}$$

(ii) Find the general solution for a rectangular waveguide (Fig. 5.7) where the boundary conditions $E_x = 0$ at $y = \pm a/2$ and $E_y = 0$ at $x = \pm b/2$ are to be satisfied.

(iii) Which are the permissible values of β_x and β_y?

(iv) What is the cut-off frequency of the TE_{67} mode?

5.10. The field components of the TE_{10} mode in a rectangular waveguide are given by eqns (5.125), (5.130), and (5.131).

(i) Determine the Poynting vector at an arbitrary point in the waveguide,

(ii) Show that the total power propagating along the waveguide may be written in the form

$$P=\frac{ab}{4}\frac{\beta}{\omega\mu_0}[|E_f|^2-|E_r|^2]. \tag{5.325}$$

5.11. The walls of a physically realizable waveguide have finite conductivity and as a consequence the fields attenuate as $\exp(-\alpha z)$, similarly to the decaying waves discussed in Section 5.11. Show that α may be expressed in the form

$$\alpha=\frac{P_l}{2P} \tag{5.326}$$

where P is the total amount of power propagating in the waveguide and P_l is the power lost per unit length of waveguide.

5.12. The power propagating in the waveguide is reduced from P_1 to P_2 between two cross-sections a distance z apart. The attenuation in decibels is then defined as $10\log_{10}(P_2/P_1)$. Assuming exponential decay of the fields (as in Example 5.11) show that the attenuation in terms of db m^{-1} may be obtained in the form

$$A = 20\alpha\log_{10}\mathrm{e} \tag{5.327}$$

5.13. The losses in a waveguide may be calculated on the following basis. Take a forward travelling wave and assume that the magnetic field is correctly given by the infinite conductivity model, e.g. H_z is given by eqn (5.130) (with $E_r = 0$) for the TE_{10} mode in a rectangular waveguide (Fig. 5.7). Assume further that on the metallic surface at $y = \pm a/2$ the relationship obtained in Example 5.5 holds, that is an electric field of magnitude $Z_c H_z$ appears perpendicular to the magnetic field. Hence the Poynting vector is directed towards the walls and its magnitude is given as

$$P_d = \tfrac{1}{2}\,\mathrm{Re}[Z_c|H_z(\pm a/2)|^2]. \tag{5.328}$$

(i) Show that the power moving into the sidewalls per unit length is

$$P_{l1} = |E_f|^2 \frac{b}{\sqrt{2}}\left(\frac{\pi}{a}\right)^2 (\omega\mu_0)^{-\frac{3}{2}}\sigma^{-\frac{1}{2}}, \tag{5.329}$$

(ii) Consider now the magnetic field at the bottom and top surfaces. Show that the power moving into the walls per unit length is

$$P_{l2} = |E_f|^2 \frac{a}{2}\left(\frac{\omega}{2\sigma}\right)^{\frac{1}{2}} \mu_0^{-\frac{3}{2}} c^{-2}, \tag{5.330}$$

(iii) Show with the aid of eqns (5.326), (5.327), (5.329) and (5.330) that the attenuation in db m^{-1} is given by the formula

$$A = 20(\log_{10} e)\left[\frac{1}{\mu_0 c\sigma a[(f/f_c)^2 - 1]}\right]^{\frac{1}{2}}\left[\frac{1}{2b}\left(\frac{f}{f_c}\right)^{\frac{3}{2}} + \frac{1}{a}\left(\frac{f}{f_c}\right)^{-\frac{1}{2}}\right], \tag{5.331}$$

(iv) What is the attenuation at $f = 10$ GHz for a copper waveguide having the dimensions $a = 25$ mm, $b = 12{\cdot}5$ mm?

(v) What is the smallest attenuation achievable in the above waveguide? At what frequency will it occur?

5.14. The electric field distribution of a TE_{01} mode in a circular waveguide is given by eqn (5.149) and the propagation coefficient by eqn (5.148).

(i) Determine the cut-off frequency, f_c.

(ii) Show that the corresponding magnetic field is given by

$$H_R = -\frac{\beta}{\omega\mu_0} J_1(\zeta R)[E_f\, e^{-j\beta z} - E_r\, e^{j\beta z}] \tag{5.332}$$

$$H_z = -\frac{\zeta}{j\omega\mu_0} J_0(\zeta R)[E_f\, e^{-j\beta z} + E_r\, e^{j\beta z}], \tag{5.333}$$

(iii) Show that the power propagating in a lossless waveguide may be written in the form

$$P = \frac{\pi\beta}{\omega\mu_0}\frac{R_0^2}{2} J_0^2(\zeta R_0)[|E_f|^2 - |E_r|^2]. \tag{5.334}$$

(Note the following relationships, valid for Bessel functions:

$$\frac{d}{dR}[RJ_1(\zeta R)] = \zeta R J_0(\zeta R), \tag{5.335}$$

$$\int_0^{R_0} RJ_1^2(\zeta R)\,dR = \frac{R_0^2}{2} J_0^2(\zeta R_0) \quad \text{when} \quad J_1(\zeta R_0) = 0.) \tag{5.336}$$

5.15. (i) Show in the manner outlined in Example 5.13 that the attenuation in db m^{-1} of the TE_{01} mode in a circular waveguide is given by the formula

$$A = 20(\log_{10} e)\left(\frac{3\cdot 832}{2\sigma\mu_0 c}\right)^{\frac{1}{2}} \frac{1}{R_0^{3/2}} \frac{(f/f_c)^{-\frac{1}{2}}}{\sqrt{\{(f/f_c)^2-1\}}}. \tag{5.337}$$

(ii) Show that the attenuation decreases monotonically as a function of frequency.

5.16. A circular waveguide of 12 cm diameter made of copper and carrying the TE_{01} mode is to be used for long distance communications. If the attenuation between two repeater stations is not to exceed 30 db what is the maximum permissible distance between the repeaters (considering attenuation alone) at a frequency of 37·5 GHz?

5.17. A length of rectangular waveguide is short-circuited at both ends (at $z = 0$ and at $z = l$) in order to make a cavity resonator (Fig. 5.15). Assume that a TE_{10} mode is excited and

(i) Show that at the resonant frequency f_r the electric and magnetic fields vary as follows

$$\begin{aligned} E_x &= A \cos\frac{\pi y}{a} \sin\frac{n\pi}{l}z \\ H_y &= -jA\frac{n}{l}\frac{1}{2f_r\mu_0}\cos\frac{\pi y}{a}\cos\frac{n\pi}{l}z \\ H_z &= jA\frac{1}{2f_r\mu_0 a}\sin\frac{\pi y}{a}\sin\frac{n\pi}{l}z \end{aligned} \tag{5.338}$$

where

$$f_r = \frac{c}{2}\left[\left(\frac{n}{l}\right)^2 + \frac{1}{a^2}\right]^{\frac{1}{2}}, \qquad A = 2jE_f = -2jE_r \tag{5.339}$$

and n is a positive integer,

(ii) Calculate the lowest resonant frequency for a cavity of $a = 25$ mm, $b =$ 12·5 mm, and $l = 30$ mm.

(iii) Show that under resonant conditions equal amount of energy is stored in the electric and magnetic fields.

5.18. A length of circular waveguide carrying the TE_{01} mode is short-circuited at $z = 0$ and at $z = l$. Show that at the resonant frequency f_r the electric and magnetic fields vary as follows

$$\begin{aligned} E_\varphi &= AJ_1(\zeta R)\sin\frac{n\pi}{l}z \\ H_R &= j\frac{n}{2lf_r\mu_0}AJ_1(\zeta R)\cos\frac{n\pi}{l}z \\ H_z &= j\frac{\zeta A}{2\pi f_r\mu_0}J_0(\zeta R)\sin\frac{n\pi}{l}z \end{aligned} \tag{5.340}$$

where

$$f_r = \frac{c}{2\pi}\left[\zeta^2 + \left(\frac{n\pi}{l}\right)^2\right]^{\frac{1}{2}}, \qquad A = 2jE_f = -2jE_r. \tag{5.341}$$

For what value of n is the resonant frequency of the TE_{01n} mode nearest to 37·5 GHz for resonator dimensions $R_0 = 0\cdot 06$ m and $l = 2$ m?

5.19. It is true in general (Example 5.17 was concerned with a special case) that at resonance the energy stored in the electric field is equal to that in the magnetic field. Note that time averaging brings in a factor $\frac{1}{2}$, hence the correct formula for the energy stored in the electric field is

$$W_{\mathrm{E}} = \tfrac{1}{4}\varepsilon \int |E|^2 \,\mathrm{d}\tau. \tag{5.342}$$

Show that the total energy stored in the cylindrical resonator of Example 5.18 is equal to

$$W = 2W_{\mathrm{E}} = \frac{\pi}{2}|A|^2 \varepsilon l R_0^2 J_0^2(\zeta R_0). \tag{5.343}$$

5.20. Show that the energy dissipated per cycle in the cylindrical resonator of Example 5.18 is equal to

$$\tfrac{1}{8}(\pi\sigma)^{-\frac{1}{2}}\mu^{-\frac{3}{2}}f_{\mathrm{r}}^{-\frac{5}{2}}|A|^2 J_0^2(\zeta R_0) R_0\left(l\zeta^2 + \frac{2R_0 n^2\pi^2}{l^2}\right). \tag{5.344}$$

(Hint: Use the method outlined in Example 5.13 for determining the losses but remember the endplates.)

5.21. Show that the Q factor (defined by eqn (5.157)) of the cylindrical resonator of Example 5.18 may be written in the form

$$Q = \frac{8\pi^{\frac{5}{2}}\varepsilon_0 \mu_0^{\frac{3}{2}} f_r^{\frac{5}{2}} l^3 R_0 \sigma^{\frac{1}{2}}}{l^3\zeta^2 + 2R_0 n^2 \pi^2}. \tag{5.345}$$

Calculate the Q factor for resonator dimensions $R_0 = 0{\cdot}06$ m and $l = 2$ m for the TE_{01n} mode resonating near to 37·5 GHz.

5.22. An air-filled rectangular waveguide resonator of dimensions $a = 25$ mm, $b = 12{\cdot}5$ mm, and $l = 60$ mm resonates in the TE_{102} mode. It will resonate at the same frequency in the TE_{103} mode if filled with a dielectric. Determine the required value of ε_{r}.

5.23. Take an open resonator (Fig. 5.17a) of length $l = 50$ mm resonating at a frequency of 5×10^{14} Hz. Since the dimensions of the resonator are very large in comparison with the wavelength one can take the wavelength as being equal to the free space wavelength. Determine approximately the number of half wavelengths in the resonator. The next mode will have one more half wavelength and hence a different resonant frequency. Determine the frequency difference between these two modes.

5.24. Take a plane wave incident perpendicularly ($\alpha_1 = 0$) from medium 1 upon medium 2 as shown in Fig. 5.19. Considering both polarizations show that the reflection coefficients are as follows

$$\Gamma_E = -\Gamma_H = \frac{Z_2 - Z_1}{Z_2 + Z_1} \tag{5.346}$$

where

$$Z_1 = \left(\frac{\mu}{\varepsilon_1}\right)^{\frac{1}{2}}, \qquad Z_2 = \left(\frac{\mu}{\varepsilon_2}\right)^{\frac{1}{2}}. \tag{5.347}$$

Why is there a sign difference between Γ_E and Γ_H?

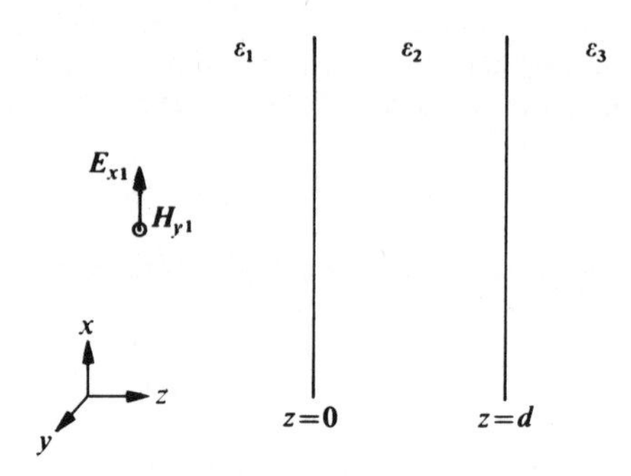

FIG. 5.44. Three lossless dielectrics.

5.25. Assume three lossless media of dielectric constants ε_1, ε_2, and ε_3 respectively as shown in Fig. 5.44. The electric and magnetic fields in a plane wave propagating in the $+z$ direction may be described as

$$E_{x1} = E_{f1} \exp(-j\beta_{m1}z) + E_{r1} \exp j\beta_{m1}z \tag{5.348}$$

and

$$H_{y1} = \frac{1}{Z_1}[E_{f1} \exp(-j\beta_{m1}z) - E_{r1} \exp j\beta_{m1}z]. \tag{5.349}$$

(i) Show that

$$\Gamma = \frac{E_{r1}}{E_{f1}} = -\frac{Z_2(Z_1 - Z_3) \cos \beta_{m2}d + j(Z_1Z_3 - Z_2^2) \sin \beta_{m2}d}{Z_2(Z_1 + Z_3) \cos \beta_{m2}d - j(Z_1Z_3 + Z_2^2) \sin \beta_{m2}d} \tag{5.350}$$

and

$$\frac{E_{x1}(0)}{H_{y1}(0)} = Z_2\frac{Z_3 \cos \beta_{m2}d + jZ_2 \sin \beta_{m2}d}{Z_2 \cos \beta_{m2}d + jZ_3 \sin \beta_{m2}d} \tag{5.351}$$

where

$$\beta_{m1} = \omega(\mu\varepsilon_1)^{\frac{1}{2}}, \qquad \beta_{m2} = \omega(\mu\varepsilon_2)^{\frac{1}{2}}, \tag{5.352}$$

d is the thickness of medium 2, and Z_1, Z_2, and Z_3 are as defined in Example 5.24.

(ii) Find the conditions under which $\Gamma = 0$. In what practical applications is it an advantage to reduce reflections to zero?

(iii) Take the thickness of medium 2 equal to quarter wavelength (measured in the medium) and find the maximum achievable reflection if medium 1 is free space and the dielectrics available for mediums 2 and 3 may have relative permittivities between 1 and 16.

5.26. Fermat's principle states that light rays moving between two points will choose the path minimizing the time of travel. Determine on this basis the laws of refraction at a dielectric interface.

5.27. For a dipole of finite length the vector potential is given by eqn (5.237). If $r_0 \gg l$ then r may be taken equal to r_0 in the denominator whereas in the exponent the approximation (see Fig. 5.30) $r = r_0 - z \cos \theta$ may be used. Take the length of the dipole as half wavelength in which case (Fig. 5.29d) the current distribution is $I_m \cos \beta_0 z$ in the region $-l/2 < z < l/2$.

(i) Show that the vector potential at the point r_0, θ takes the form

$$A_z = \frac{I_m \exp(-j\beta_o r_0)}{2\pi\beta r_0} \frac{\cos(\pi/2 \cos\theta)}{\sin^2\theta}. \tag{5.353}$$

(ii) Determine first the magnetic and then the electric field (retain only the $1/r_0$ terms).

(iii) Plot the radiation pattern in polar coordinates.

(iv) Determine the Poynting vector.

(v) Determine the radiation resistance

(Note that

$$\int_0^{\pi/2} \frac{\cos^2(\pi/2 \cos\theta)}{\sin\theta} d\theta = 0{\cdot}609 \tag{5.354}$$

as obtained by numerical techniques).

(vi) Determine the gain.

5.28. A dielectric lens of diameter D is placed at a distance f from a point radiator. Determine the profile of the lens if the aim is to obtain a plane wavefront. Assume that $D \gg \lambda$ so that the laws of geometrical optics are applicable (Fig. 5.45).

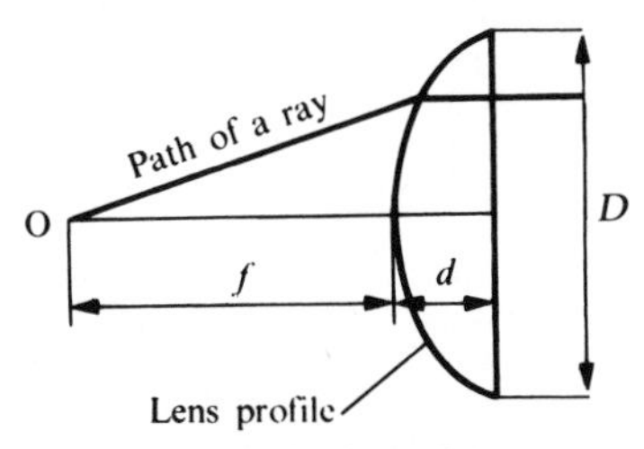

FIG. 5.45. A dielectric lens.

5.29. Prove that eqns (5.279) to (5.281) satisfy eqn (5.277).

5.30. (i) Determine the radiation patterns both in the xz and yz planes (Fig. 5.34) when the respective field distributions are (see eqn (5.264)) as follows

$$f_1(x) = 1, \qquad f_2(y) = \cos\frac{\pi y}{b}. \tag{5.355}$$

(ii) Plot the radiation patterns as functions of the parameters

$$u = \frac{\beta_0 a}{2} \sin\theta \quad \text{and} \quad v = \frac{\beta_0 b}{2} \sin\theta \tag{5.356}$$

respectively.

(iii) Calculate the gain.

(iv) How does the gain compare to that of a uniformly illuminated aperture?

5.31. The field distribution in the x direction of an aperture (Fig. 5.34) is given by the function

$$f_1(x) = \exp\left(-j\frac{2\alpha x}{a}\right), \tag{5.357}$$

that is the amplitude is constant but the phase varies linearly across the aperture.

(i) Determine the radiation pattern in the xz plane.

(ii) What is the direction of maximum radiation if $\alpha \ll 1$ and $\beta_0 a \gg 1$?

6. Relativity

Are not gross Bodies and Light convertible into another, and may not Bodies receive much of their Activity from the Particles of Light which enter their Composition?

The changing of Bodies into Light, and Light into Bodies, is very conformable to the Course of Nature, which seems delighted with Transmutations.

ISAAC NEWTON *Opticks* 2nd edition, 1717

Man denke z. B. an die elektrodynamische Wechselwirkung zwischen einem Magneten und einem Leiter. Das beobachtbare Phänomen hängt hier nur ab von der Relativbewegung von Leiter und Magnet, während nach der üblichen Auffassung die beiden Fälle, daß der eine oder der andere dieser Körper der bewegte sei, streng voneinander zu trennen sind. Bewegt sich nämlich der Magnet und ruht der Leiter, so entsteht in der Umgebung des Magneten ein elektrisches Feld von gewissem Energiewerte, welches an den Orten, wo sich Teile des Leiters befinden, einen Strom erzeugt. Ruht aber der Magnet und bewegt sich der Leiter, so entsteht in der Umgebung des Magneten kein elektrisches Feld, dagegen im Leiter eine elektromotorische Kraft, welcher an sich keine Energie entspricht, die aber—Gleichheit der Relativbewegung bei den beiden ins Auge gefaßten Fällen vorausgesetzt—zu elektrischen Strömen von derselben Größe und demselben Verlaufe Veranlassung gibt, wie im ersten Falle die elektrischen Kräfte.

Beispiele ähnlicher Art, sowie die mißlungenen Versuche, eine Bewegung der Erde relativ zum 'Lichtmedium' zu konstatieren, führen zu der Vermutung, daß dem Begriffe der absoluten Ruhe nicht nur in der Mechanik, sondern auch in der Elektrodynamik keine Eigenschaften der Erscheinungen entsprechen.

Die Einführung eines 'Lichtäthers' wird sich insofern als überflüssig erweisen, als nach der zu entwickelnden Auffassung weder ein mit besonderen Eigenschaften ausgestatteter 'absolut ruhender Raum' eingeführt.

ALBERT EINSTEIN *Zur Elektrodynamik bewegter Körper, Annalen der Physik* **17** 1905

Gibt ein Körper die Energie L in Form von Strahlung ab, so verkleinert sich seine Masse um L/V^2. Hierbei ist es offenbar unwesentlich, daß die dem Körper entzogene Energie gerade in Energie der Strahlung übergeht, so daß wir zu der allgemeineren Folgerung geführt werden:

Die Masse eines Körpers ist ein Maß für dessen Energieinhalt; ändert sich die Energie um L, so ändert sich die Masse in demselben Sinne um $L/9 \cdot 10^{20}$, wenn die Energie in Erg und die Masse in Grammen gemessen wird.

Es ist nicht ausgeschlossen, daß bei Körpen, deren Energieinhalt in hohem Maße veränderlich ist (z. B. bei den Radiumsalzen), eine Prüfung der Theorie gelingen wird.

Wenn die Theorie den Tatsachen entspricht, so überträgt die Strahlung Trägheit zwischen den emittierenden und absorbierenden Körpern.

ALBERT EINSTEIN *Ist die Trägheit eines Körpers von seinem Energieeinhalt abhängig? Annalen der Physik* **18** 1905

6.1. Introduction

How much relativity should an engineer know? Most textbooks written for engineers omit relativity altogether.† Is that justified? I can see arguments for it and against it. Those against are as follows.

1. The design equations for producing high-velocity (near to the velocity of light) particle beams are based on the theory of relativity.

2. The basic equation, $W=mc^2$, making possible the exploitation of nuclear energy, comes from the theory of relativity.‡

3. Whenever relative motion is considered (expecting, for example, the same effect whether a wire loop moves in a stationary magnetic field or the magnetic field sweeps past a stationary loop) one is in fact using the theory of relativity.

4. Any engineer should be familiar with the basic theories of physics; and relativity is one of them.

The arguments in favour of omitting it are no less convincing.

1. Very few engineers design high-velocity particle beams. The modifications introduced by relativity have profound effects upon the behaviour of the beam but the design equations don't become unduly difficult. If you ever get a job like that you'll have plenty of time to look up the design equations.

2. Engineers are quite willing to accept a simple formula like $W=mc^2$, without the need to enquire into its origin. Unfamiliarity with the theory of relativity is the least of the troubles of the nuclear engineer. He needs to worry much more about the strength of materials, control, safety, etc., all traditional engineering problems.

3. When relative motion is considered by heavy-current engineers they use common sense and not the theory of relativity. Light-current engineers, on the other hand, have not developed the taste for adopting moving coordinate systems. I spent a number of years analysing the interaction of charged beams with travelling electromagnetic waves, but it never occurred

† With a few notable exceptions, such as G. W. Carter (1954). *The electromagnetic field in its engineering aspects.* Longmans, Green, and Co., London and K. Simonyi (1963). *Foundations of electrical engineering.* Pergamon Press, Oxford; and there is a book on relativity written specifically for engineers with the induction effect in mind, E. G. Cullwick (1957). *Electromagnetism and relativity.* Longmans, Green, and Co., London.

‡ I am using here W for energy, with the greatest reluctance. The form in which this equation has acquired fame and recognition is $E=mc^2$. Unfortunately, the symbol E is reserved for the electric field in our notation.

to me to look at the electromagnetic waves from the point of view of the observer travelling at the velocity of the charged beam. It's not done. All the literature of the subject is written in the laboratory frame of reference. The reason might be that we are afraid of losing our painfully acquired physical intuition. We have learned a lot about the behaviour of electric and magnetic fields in the laboratory frame of reference and might be reluctant to concede that the electric and magnetic fields look different to observers moving at different speeds.

4. Theories of physics are getting so complicated that an engineer must draw the line somewhere. There isn't enough time to acquire familiarity with all the basic theories of physics. If relativity does not help in the solution of engineering problems there is no need to know about it.

The unimportance of relativity to engineers is even more apparent if we compare it with that other newcomer to engineering problems, quantum mechanics. If some powerful god would command tomorrow that 'thou shalt not have quantum mechanics', it would cause havoc in electrical engineering (quite apart from the fact that atoms would fall apart). A similar command concerning relativity would leave engineers indifferent.

As you can see the arguments are neatly balanced. I have no good reasons for coming down on the side of relativity. It might be a mistaken move, but since the effort required is relatively small, you have little to lose.

6.2. The basic assumptions

Relativity is one of the subjects everyone knows about. Even arts graduates can tell you that the theory was first propounded by Albert Einstein and that it is of the utmost significance. There are many fewer people who can tell you the basic assumptions, although they can be stated in surprisingly simple terms.

1. The laws of physics are the same in all coordinate systems which move uniformly relative to each other.
2. The velocity of light in empty space is the same in all reference frames and is independent of the motion of the emitting body.
3. $W = mc^2$. There are many ways of deriving this relationship, showing the equivalence of mass and energy, from the two previous assumptions. For our (not very rigorous) purpose it will be perfectly adequate to regard it as a separate assumption.

6.3. Velocity-dependent mass

You have made lots of calculations in mechanics with the aid of Newton's equation, claiming that mass times acceleration is equal to force. In the usual mathematical form,

$$m\frac{d\mathbf{v}}{dt} = \mathbf{F}. \tag{6.1}$$

We shall now abandon the assumption that mass is constant† and rewrite eqn (6.1) in the slightly different form:

$$\frac{\mathrm{d}}{\mathrm{d}t}(m\mathbf{v}) = \mathbf{F}. \tag{6.2}$$

You may also remember that power, the rate of change of energy, is equal to force times velocity:

$$\frac{\mathrm{d}W}{\mathrm{d}t} = \mathbf{F}.\mathbf{v}, \tag{6.3}$$

which becomes now

$$\frac{\mathrm{d}}{\mathrm{d}t}(mc^2) = \mathbf{v}\,.\,\frac{\mathrm{d}}{\mathrm{d}t}(m\mathbf{v}). \tag{6.4}$$

Multiplying eqn (6.4) by $2m$ and integrating we get

$$m^2c^2 = m^2\mathbf{v}^2 + C. \tag{6.5}$$

The constant C may be determined by assigning the mass m_0 (rest mass) to the case when $v = 0$ leading to

$$C = m_0^2c^2, \tag{6.6}$$

which, after substitution into eqn (6.5), yields for the mass

$$m = m_0\gamma, \qquad \gamma = \left(1 - \frac{v^2}{c^2}\right)^{-\frac{1}{2}}. \tag{6.7}$$

It may be seen that $\gamma \simeq 1$ when $v \ll c$, and mass may be regarded a constant, but in general mass is velocity-dependent, tending to infinity as v approaches the velocity of light.

6.4. The relativity of simultaneity and of length

It is rather disturbing that mass is velocity-dependent because it undermines our belief in the constancy of things (if mass isn't constant, what is?) but there are worse surprises to come. It turns out that no unambiguous answer is possible even to such a simple question as 'when did a particular event occur?' The example often quoted relies on stationary observers at railway platforms and observers travelling on a train with uniform speed. In 1974 we should be more modern and imagine instead a geostationary space station and a spaceship passing it. The stationmaster and the instruments with which he can make precise measurements are situated at S_0 (Fig. 6.1(a)). He sends out two of his minions with light-flash equipments and synchronized clocks to points S_1 and S_2 such that $S_0S_1 = S_0S_2$. At an agreed

† Newton worked on the assumption that mass was constant, but interestingly when formulating his first law he wrote it in a more general form referring to the change in '*motus*' (*mutatio motus proportionalis est*...), which we call nowadays momentum.

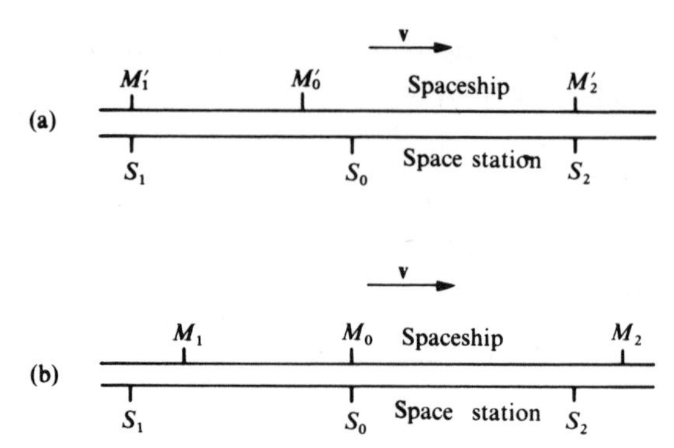

FIG. 6.1. Schematic drawing of a spaceship passing a space station. Positions of the observers at the time: (a) when light flashes are sent out at S_1 and S_2; (b) when light flashes arrive at S_0.

time both men send out light flashes which will be received *simultaneously* at S_0. So far there is no problem. Provided that (1) $S_0S_1 = S_0S_2$, (2) the flashes are sent at the same time, and (3) light sent from S_2 propagates with the same velocity as light sent from S_1, the two flashes are bound to arrive at the same time at S_0.

Let it happen that a spaceship passes by (at uniform speed of course) just at the time when the stationmaster and his men play with their light flashes. Let us further assume that midshipman Joe Blogg is just opposite the stationmaster at point M_0 when the flashes arrive. There are also two attentive 'edgeshipmen' on the spaceship who happen to be at points M_1' and M_2' (opposite to S_1 and S_2) when the flashes occur. This is shown in Fig. 6.1(a). Note that at the time the flashes occur Joe Blogg is not yet opposite to the stationmaster. Assuming that the spaceship is moving to the right, he will be at the point M_0'. By the time the stationmaster receives the light flashes Midshipman Joe Blogg will be at M_0 and the two edgeshipmen will be at M_1 and M_2 respectively (Fig. 6.1(b)). As agreed, Joe Blogg will receive the light flashes at the same time as the stationmaster.

Suppose that Joe Blogg wants to find out when the flashes originated. He knows (or can find out) that his friends were at M_1' and at M_2' at the time. He can also measure the distances $M_0'M_2' = M_0M_2$ and $M_0'M_1' = M_0M_1$ on the spaceship and find that $M_0M_1 < M_0M_2$. Hence he will conclude that the light flash from M_2 was initiated at an earlier time than that from M_1. This is a straightforward conclusion on his part. He assumes that light travels with a velocity c inside the spaceship, and light coming from the left travels with the same velocity as light coming from the right. This is in accordance with postulate (2) of Section 6.2, which states that the velocity of light inside the spaceship is independent of the velocity of the spaceship.

The stationmaster concludes that the flashes originated at the same time: Joe Blogg maintains that one occurred earlier than the other one. Who is right? One can't tell. The emerging conclusion is that there is no such thing as simultaneity of two events. Simultaneity is relative; it depends on the speed of the observer.

What about length? Let's measure length by the following method (Fig. 6.2). Light flashes are sent simultaneously from points A and B. The stationary observer situated at O can work out the length $l = AB$ from the

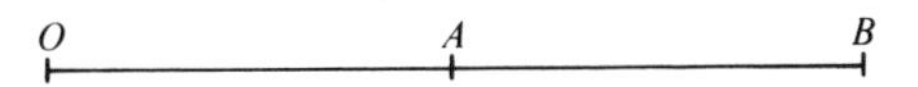

FIG. 6.2. Observer positioned at O measures the length AB by measuring the time difference of arrival of light flashes sent out simultaneously from A and B.

time difference at which the flashes arrive. Unfortunately, observers moving at different speeds cannot agree on simultaneity, hence they cannot agree on length either. It comes as a shattering blow to our commonsense approach to the world but we must accept it; there is no absolute time and there is no absolute length.

6.5. The Lorentz transformations

We have seen that two observers moving at a uniform velocity relative to each other have different time and length scales. We shall now be concerned with the transformation of variables from one coordinate system into another, from the stationary coordinate system of the stationmaster into the moving coordinate system of Joe Blogg. Let us start by putting them opposite to each other (as in Fig. 6.1(b)) at $t = 0$ and send out a light flash from the same point at the same time. How far will the light pulse move in each coordinate system? In the stationmaster's coordinate system (x, y, z, t) after a time t the light arrives at the surface of a sphere for which

$$x^2 + y^2 + z^2 = c^2t^2. \tag{6.8}$$

This is obvious. All we have said is that light propagates with a velocity c in all directions. In view of postulate (2) the same is true in the coordinate system (x', y', z', t') of Joe Blogg. Looking from the spaceship, light propagates with the same velocity in all directions, hence after a time t' it will have reached the surface of the sphere

$$x'^2 + y'^2 + z'^2 = c^2t'^2. \tag{6.9a}$$

Now at least one rule of the coordinate transformation is clear; the transformation must leave the quantity

$$x^2 + y^2 + z^2 - c^2t^2 \tag{6.9b}$$

invariant. Luckily, there is only one transformation that can satisfy the above conditions without having any singularities, the so-called Lorentz transformation. It takes the form (see Example 6.6)

$$x' = \gamma(x - vt), \qquad y' = y, \qquad z' = z, \qquad t' = \gamma\left(t - \frac{vx}{c^2}\right), \tag{6.10a}$$

where the x', y', z' three-dimensional coordinate system travels with a velocity v in the x' direction relative to the x, y, z stationary coordinate system. In the future I shall refer to these coordinate systems as the S and S' systems (or frames) respectively. The inverse transformation is similarly given by

$$x = \gamma(x' + vt'), \qquad y = y', \qquad z = z', \qquad t = \gamma\left(t' + \frac{vx'}{c^2}\right). \tag{6.10b}$$

Let us put now into mathematical form some of the loose talk of the previous section. Take a rod of length l' that is at rest in S' and lies along the x' axis. Its length is determined by the coordinates of the two ends, namely

$$l' = x_2' - x_1'. \tag{6.11a}$$

As mentioned before the observer in S has to perform the measurement at the same time t according to *his* time scale. He takes the length as the difference between his coordinates x_2 and x_1, i.e.

$$l = x_2 - x_1 = \frac{x_2'}{\gamma} + vt - \left(\frac{x_1'}{\gamma} + vt\right) = \frac{l'}{\gamma}. \tag{6.11b}$$

This is the so called Lorentz–Fitzgerald contraction.† Whenever a body is in motion relative to the observer its measured length is contracted in the direction of its motion by the factor γ. Next we shall compare times. If two events occur in the S' system at the coordinate x_0' at times t_1' and t_2' then the same time interval in the S system will be measured as

$$T = t_2 - t_1 = \gamma\left(t_2' + \frac{vx_0'}{c^2}\right) - \gamma\left(t_1' + \frac{vx_0'}{c^2}\right) = \gamma(t_2' - t_1')$$
$$= \gamma T'. \tag{6.12}$$

A time interval as measured by a clock at rest in S' is smaller than that measured in S. As a consequence an astronaut travelling at high speeds and away for a time interval T' according to his clocks will find that a time $\gamma T'$ has passed meanwhile on Earth.‡

† A somewhat unexpected consequence of this contraction is discussed in the following limerick.

There was a young man called Fiske,
Whose fencing was strikingly brisk,
So fast was his action
Fitzgerald contraction
Reduced his rapier to a disc.

‡ As far as space travellers are concerned there is no experimental proof as yet so we must be content to accept evidence concerning the lifetime of μ-mesons. As measured by Rossi and Hall in 1941 (*Phys. Rev.* **59** 223) those travelling with a velocity $v \simeq 0{\cdot}994c$ live about 9 times longer than those stationary in the laboratory frame.

One can have great fun considering the various implications of the Lorentz transformation. Take for example a particle that moves with u' in the x', y', z', t' coordinate system which itself moves with velocity v relative to the x, y, z, t coordinate system. What is the velocity of that particle in the latter coordinate system? An interesting problem you should be able to solve, see Example 6.7.

If the laws of nature are the same in all coordinate systems moving uniformly relative to each other then the Lorentz transformation should leave Maxwell's equations invariant. What does it mean: 'leaving Maxwell's equations invariant'? It means that they look the same in the x', y', z', t' coordinate system as in the x, y, z, t coordinate system.

We know how Maxwell's equations look in the x, y, z, t system. I have written those equations umpteen times. Let us just write two of them for the simple case of a vacuum, namely:

$$\nabla\times\mathbf{B}=\frac{1}{c^2}\frac{\partial\mathbf{E}}{\partial t}+\mu_0\mathbf{J}\quad\text{and}\quad\nabla\cdot\mathbf{E}=\frac{\rho}{\varepsilon_0}. \tag{6.13}$$

How will the above equations look in the x', y', z', t' system? Perhaps

$$\nabla'\times\mathbf{B}=\frac{1}{c^2}\frac{\partial\mathbf{E}}{\partial t'}+\mu_0\mathbf{J}\quad\text{and}\quad\nabla'\cdot\mathbf{E}=\frac{\rho}{\varepsilon_0}. \tag{6.14}$$

No, that's definitely wrong. The electric and magnetic fields do not preserve their values in a Lorentz transformation. The correct form is

$$\nabla'\times\mathbf{B}'=\frac{1}{c^2}\frac{\partial\mathbf{E}'}{\partial t}+\mu_0\mathbf{J}'\quad\text{and}\quad\nabla'\cdot\mathbf{E}'=\frac{\rho'}{\varepsilon_0}. \tag{6.15}$$

In order to find $\mathbf{B}'$, $\mathbf{E}'$, and $\mathbf{J}'$ in terms of $\mathbf{B}$, $\mathbf{E}$, and $\mathbf{J}$ we shall have to perform the transformation of the coordinates according to eqn (6.10a). We shall need

$$\frac{\partial}{\partial t}=\frac{\partial}{\partial x'}\frac{\partial x'}{\partial t}+\frac{\partial}{\partial t'}\frac{\partial t'}{\partial t}=\gamma\left(-v\frac{\partial}{\partial x'}+\frac{\partial}{\partial t'}\right) \tag{6.16}$$

and

$$\begin{aligned}\frac{\partial}{\partial x}&=\frac{\partial}{\partial x'}\frac{\partial x'}{\partial x}+\frac{\partial}{\partial t'}\frac{\partial t'}{\partial x}=\gamma\left(\frac{\partial}{\partial x'}-\frac{v}{c^2}\frac{\partial}{\partial t'}\right),\\ \frac{\partial}{\partial y}&=\frac{\partial}{\partial y'},\\ \frac{\partial}{\partial z}&=\frac{\partial}{\partial z'}.\end{aligned} \tag{6.17}$$

The equations to be transformed are as follows:

$$\frac{\partial B_x}{\partial z}-\frac{\partial B_z}{\partial x}=\frac{1}{c^2}\frac{\partial E_y}{\partial t}+\mu_0 J_y, \tag{6.18}$$

$$\frac{\partial B_y}{\partial x}-\frac{\partial B_x}{\partial y}=\frac{1}{c^2}\frac{\partial E_z}{\partial t}+\mu_0 J_z, \tag{6.19}$$

$$\frac{\partial B_z}{\partial y}-\frac{\partial B_y}{\partial z}=\frac{1}{c^2}\frac{\partial E_x}{\partial t}+\mu_0 J_x, \tag{6.20}$$

and

$$\frac{\partial E_x}{\partial x}+\frac{\partial E_y}{\partial y}+\frac{\partial E_z}{\partial z}=\frac{\rho}{\varepsilon_0}. \tag{6.21}$$

Eqn (6.18) becomes after transformation

$$\frac{\partial B_x}{\partial z'}-\gamma\left(\frac{\partial}{\partial x'}-\frac{v}{c^2}\frac{\partial}{\partial t'}\right)B_z=\frac{\gamma}{c^2}\left(-v\frac{\partial}{\partial x'}+\frac{\partial}{\partial t'}\right)E_y+\mu_0 J_y, \tag{6.22}$$

which may be rewritten as

$$\frac{\partial B_x}{\partial z'}-\frac{\partial}{\partial x'}\left\{\gamma\left(B_z-\frac{v}{c^2}E_y\right)\right\}=\frac{1}{c^2}\frac{\partial}{\partial t'}\{\gamma(E_y-vB_z)\}+\mu_0 J_y. \tag{6.23}$$

After the first transformation we can already see some rules of the game. Eqn (6.23) will have the same form as eqn (6.18) provided

$$B'_x=B_x,\qquad B'_z=\gamma\left(B_z-\frac{v}{c^2}E_y\right),$$
$$E'_y=\gamma(E_y-vB_z),\qquad J'_y=J_y. \tag{6.24}$$

Let us transform now eqn (6.19). We get

$$\gamma\left(\frac{\partial}{\partial x'}-\frac{v}{c^2}\frac{\partial}{\partial t'}\right)B_y-\frac{\partial B_x}{\partial y'}=\frac{\gamma}{c^2}\left(-v\frac{\partial}{\partial x'}+\frac{\partial}{\partial t'}\right)E_z+\mu_0 J_z, \tag{6.25}$$

which may be rewritten as

$$\frac{\partial}{\partial x'}\left\{\gamma\left(B_y+\frac{v}{c^2}E_z\right)\right\}-\frac{\partial B_x}{\partial y'}=\frac{1}{c^2}\frac{\partial}{\partial t'}\{\gamma(E_z+vB_y)\}+\mu_0 J_z, \tag{6.26}$$

yielding

$$B'_y=\gamma\left(B_y+\frac{v}{c^2}E_z\right),\qquad E'_z=\gamma(E_z+vB_y),\qquad J'_z=J_z. \tag{6.27}$$

The transformation of eqn (6.20) is not less straightforward:

$$\frac{\partial B_z}{\partial y'}-\frac{\partial B_y}{\partial z'}=\frac{\gamma}{c^2}\left(-v\frac{\partial}{\partial x'}+\frac{\partial}{\partial t'}\right)E_x+\mu_0 J_x, \tag{6.28}$$

but in order to bring it to a suitable form we need first the transform of eqn (6.21) which is of the form

$$\gamma\left(\frac{\partial}{\partial x'}-\frac{v}{c^2}\frac{\partial}{\partial t'}\right)E_x+\frac{\partial E_y}{\partial y'}+\frac{\partial E_z}{\partial z'}=\frac{\rho}{\varepsilon_0}. \tag{6.29}$$

Expressing $\gamma(\partial E_x/\partial x')$ from eqn (6.29), substituting it into eqn (6.28) and rearranging it we get

$$\frac{\partial}{\partial y'}\left\{\gamma\left(B_z-\frac{v}{c^2}E_y\right)\right\}-\frac{\partial}{\partial z'}\left\{\gamma\left(B_y+\frac{v}{c^2}E_z\right)\right\}=\frac{1}{c^2}\frac{\partial E_x}{\partial t'}+\mu_0\gamma(J_x-v\rho), \tag{6.30}$$

yielding

$$E'_x=E_x, \qquad J'_x=\gamma(J_x-v\rho). \tag{6.31}$$

Expressing $(\gamma/c^2)\partial E_x/\partial t'$ from eqn (6.28) and substituting it into eqn (6.29), we get

$$\frac{\partial E_x}{\partial x'}+\frac{\partial}{\partial y'}\{\gamma(E_y-vB_z)\}+\frac{\partial}{\partial z'}\{\gamma(E_z+vB_y)\}=-\frac{1}{\varepsilon_0}\gamma\left(\rho-\frac{v}{c^2}J_x\right), \tag{6.32}$$

yielding

$$\rho'=\gamma\left(\rho-\frac{v}{c^2}J_x\right). \tag{6.33}$$

Transformation of the equations

$$\nabla\times\mathbf{E}=-\frac{\partial\mathbf{B}}{\partial t} \quad \text{and} \quad \nabla\cdot\mathbf{B}=0 \tag{6.34}$$

would just confirm what we have already found. So we know now how the electric and magnetic fields, the current density, and charge density transform. Note that the transformations may be written more compactly as

$$\begin{aligned}
&\rho'=\gamma\left(\rho-\frac{v}{c^2}J_\parallel\right)\\
&J'_\parallel=\gamma(J_\parallel-v\rho), \qquad J'_\perp=J_\perp,\\
&E'_\parallel=E_\parallel, \qquad E'_\perp=\gamma(\mathbf{E}+\mathbf{v}\times\mathbf{B})_\perp,\\
&B'_\parallel=B_\parallel, \qquad B'_\perp=\gamma\left(\mathbf{B}-\frac{1}{c^2}\mathbf{v}\times\mathbf{E}\right)_\perp,
\end{aligned} \tag{6.35}$$

where the subscripts $\parallel$ and $\perp$ denote directions parallel and perpendicular to $\mathbf{v}$.

In the the special case when $v \ll c$ the above equations simplify to

$$\rho' = \rho - \frac{1}{c^2}\mathbf{J}\cdot\mathbf{v}, \qquad \mathbf{J}' = \mathbf{J} - \rho\mathbf{v}$$

$$\mathbf{E}' = \mathbf{E} + \mathbf{v}\times\mathbf{B}, \qquad \mathbf{B}' = \mathbf{B} - \frac{1}{c^2}\mathbf{v}\times\mathbf{E}. \tag{6.36}$$

6.6. Examples

We are now in a position to give a rigorous answer to the problem raised in Section 4.6, where we were concerned with the motion of a conducting wire in a static magnetic field. In the stationary system our variables were

$$\mathbf{E} = 0,\ \mathbf{B},\ \text{and}\ \mathbf{v}. \tag{6.37}$$

Changing now to the coordinate system of the moving loop, and noting that $v \ll c$, we get from eqn (6.36)

$$\mathbf{E}' = \mathbf{v}\times\mathbf{B}, \tag{6.38}$$

the same result we arrived at in Section 4.6 when the magnetic field was assumed to travel with a velocity $\mathbf{u} = -\mathbf{v}$.†

As our second example we shall consider the force on a charged particle moving with a non-relativistic velocity. This is of course one of our basic equations first written in eqn (1.7),

$$\mathbf{F} = q(\mathbf{E} + \mathbf{v}\times\mathbf{B}). \tag{6.39}$$

Let us change now to the reference frame of the moving particle. In the new reference frame the particle is at rest, hence the force on it is due to the electric field only

$$\mathbf{F}' = q\mathbf{E}', \tag{6.40}$$

which in view of eqn. (6.36) may also be expressed as

$$\mathbf{F}' = q\mathbf{E}' = q(\mathbf{E} + \mathbf{v}\times\mathbf{B}) = \mathbf{F}, \tag{6.41}$$

i.e. the force on the particle is the same in both coordinate systems, a result one would expect from common sense.

As our third example we shall have another look at the cylindrical electron beam of Section 3.2. We determined there the force on the outermost electron at $R = a$ obtaining the results

$$E_R = \frac{\rho_0 a}{2\varepsilon_0}, \qquad B_\varphi = \tfrac{1}{2}\mu_0\rho_0 va, \qquad F = \frac{e\rho_0 a}{2\varepsilon_0}\left(1 - \frac{v^2}{c^2}\right). \tag{6.42}$$

We are going to determine now the force in the reference frame of the moving electrons. In that reference frame the electrons are at rest so the

† Shouldn't the result derived in Section 4.6 agree with the general formulae of eqn (6.35)? No, because that derivation was based on our commonsense concept of time and not on the Lorentz transformation.

current is zero. Will our equations show this? Let's see. According to eqn (6.35) the new value of current density (in the direction of the new coordinate system) is

$$J'_{\parallel} = \gamma(J_{\parallel} - v\rho). \tag{6.43}$$

In the old coordinate system

$$J_{\parallel} = v\rho, \tag{6.44}$$

hence

$$J'_{\parallel} = 0, \tag{6.45}$$

as expected.

What about the magnetic field? Well, if there is no current, there can't be magnetic field either. Certainly, but let's show this formally from our equations. Calculating first

$$\mathbf{v} \times \mathbf{E} = v\mathbf{i}_z \times \frac{a\rho_0}{2\varepsilon_0}\mathbf{i}_R = \frac{av\rho_0}{2\varepsilon_0}\mathbf{i}_\varphi, \tag{6.46}$$

we get

$$B'_{\perp} = \gamma \frac{av\rho_0}{2}\left(\mu_0 - \frac{1}{c^2\varepsilon_0}\right) = 0. \tag{6.47}$$

Now let us determine the transformed electric field. It comes to

$$E'_{\perp} = \left|\gamma\left(\frac{\rho_0 a}{2\varepsilon_0}\mathbf{i}_R + v\mathbf{i}_z + \tfrac{1}{2}\mu_0\rho_0 va\mathbf{i}_\varphi\right)\right| = \frac{\rho_0 a}{2\varepsilon_0}\left(1 - \frac{v^2}{c^2}\right)^{\frac{1}{2}}. \tag{6.48}$$

An alternative way of deriving the electric field is to start with

$$E'_R = \frac{a\rho'_0}{2\varepsilon_0}, \tag{6.49}$$

and then transform the charge density according to eqn (6.35), yielding

$$E'_R = \frac{a}{2\varepsilon_0}\gamma\left(\rho_0 - \frac{v}{c^2}v\rho_0\right) = \frac{a\rho_0}{2\varepsilon_0}\left(1 - \frac{v^2}{c^2}\right)^{\frac{1}{2}}. \tag{6.50}$$

The force in the new frame of reference is then

$$F' = eE'_R = \frac{e\rho_0 a}{2\varepsilon_0}\left(1 - \frac{v^2}{c^2}\right)^{\frac{1}{2}}. \tag{6.51}$$

The ratio of the forces comes to

$$\frac{F'}{F} = \left(1 - \frac{v^2}{c^2}\right)^{-\frac{1}{2}} = \gamma. \tag{6.52}$$

We have obtained the result that the forces in the two reference frames are not equal. Why were they found equal in the previous example? Because we made there the $v \ll c$ assumption. Under that assumption it would again be true that $F' = F$. But for relativistic speeds the forces are in general different. This follows from the definition of force as

$$\mathbf{F} = \frac{\mathrm{d}}{\mathrm{d}t}(m\mathbf{v}). \tag{6.53}$$

In the present case we are concerned with transverse momentum only which remains invariant in our transformation. The difference in force is then due to the different time scales. In the reference frame where the electron is at rest time intervals appear to be shorter by the factor γ, i.e.

$$\mathrm{d}t' = \mathrm{d}t/\gamma, \tag{6.54}$$

so force should appear larger by the same factor and that is borne out by eqn (6.52).

6.7. Four-vectors

'Four-vectors' is a technical term referring to vectors in a special type of four-dimensional space. What is the advantage of working with four-dimensional vectors? Well, what is the advantage of working with ordinary three-dimensional vectors? Simplicity and economy. Vector equations are much more compact and need no rewriting whenever we change the orientation of our coordinate system. If we denote the three rectangular coordinates for a given orientation by x_1, x_2, x_3 then the length squared of a vector

$$\mathbf{r}^2 = x_1^2 + x_2^2 + x_3^2 \tag{6.55}$$

is independent of the orientation of the coordinate system. Another example is the scalar product which will also remain invariant. With

$$\mathbf{a} = (a_1, a_2, a_3) \quad \text{and} \quad \mathbf{b} = (b_1, b_2, b_3), \tag{6.56}$$

the scalar product

$$\mathbf{a} \,.\, \mathbf{b} = a_1 b_1 + a_2 b_2 + a_3 b_3 \tag{6.57}$$

does not depend on the choice of the coordinate system.

The whole idea of introducing four-vectors is to make use of the invariance of vector formulations. We have already come across one invariant quantity and that was given by the expression

$$x^2 + y^2 + z^2 - c^2 t^2. \tag{6.9b}$$

Hence if we introduce the new coordinates

$$x_1 = x, \qquad x_2 = y, \qquad x_3 = z, \qquad x_4 = \mathrm{j}ct, \tag{6.58}$$

then the length squared

$$x_1^2+x_2^2+x_3^2+x_4^2 \tag{6.59}$$

is constant in this four-dimensional space.

The Lorentz transformation should then appear as a 'rotation' in our four-dimensional space. That is indeed so as may be seen by rewriting eqn (6.10a) in the matrix form

$$\begin{bmatrix} x_1' \\ x_2' \\ x_3' \\ x_4' \end{bmatrix} = \begin{bmatrix} \gamma & 0 & 0 & \mathrm{j}\gamma v/c \\ 0 & 1 & 0 & 0 \\ 0 & 0 & 1 & 0 \\ -\mathrm{j}\gamma v/c & 0 & 0 & \gamma \end{bmatrix} \begin{bmatrix} x_1 \\ x_2 \\ x_3 \\ x_4 \end{bmatrix} \tag{6.60}$$

exhibiting an anti-symmetric transformation matrix.

How will our equations look in the new notation? Let us first take the continuity equation

$$\nabla\cdot\mathbf{J}+\frac{\partial\rho}{\partial t}=0. \tag{5.29}$$

Introducing the notation

$$J_1=J_x, \qquad J_2=J_y, \qquad J_3=J_z, \qquad J_4=\mathrm{j}c\rho, \tag{6.61}$$

eqn (5.29) may be rewritten as

$$\frac{\partial J_1}{\partial x_1}+\frac{\partial J_2}{\partial x_2}+\frac{\partial J_3}{\partial x_3}+\frac{\partial J_4}{\partial x_4}=0. \tag{6.62}$$

Introducing further the four-dimensional vector operator

$$\tilde{\nabla}=\left(\frac{\partial}{\partial x_1},\frac{\partial}{\partial x_2},\frac{\partial}{\partial x_3},\frac{\partial}{\partial x_4}\right), \tag{6.63}$$

we get the compact form

$$\tilde{\nabla}\cdot\tilde{\mathbf{J}}=0, \tag{6.64}$$

where we have adopted the notation that a wavy bar above a bold letter means a four-dimensional vector.

Next we shall put into the new form the differential equations (5.13) and (5.15) derived for the scalar and vector potentials in vacuum,

$$\nabla^2\mathbf{A}-\frac{1}{c^2}\frac{\partial^2\mathbf{A}}{\partial t^2}=-\mu_0\mathbf{J} \tag{6.65}$$

and

$$\nabla^2\phi-\frac{1}{c^2}\frac{\partial^2\phi}{\partial t^2}=-\frac{\rho}{\varepsilon_0}. \tag{6.66}$$

Defining the four-dimensional differential operator

$$\tilde{\nabla}^2 = \nabla^2 - \frac{1}{c^2}\frac{\partial^2}{\partial t^2} = \frac{\partial^2}{\partial x_1^2} + \frac{\partial^2}{\partial x_2^2} + \frac{\partial^2}{\partial x_3^2} + \frac{\partial^2}{\partial x_4^2} \tag{6.67}$$

and introducing the four-dimensional vector

$$\tilde{\mathbf{A}} = \left(A_1, A_2, A_3, \frac{\mathrm{j}}{c}\phi\right), \tag{6.68}$$

eqns (6.64) and (6.65) may be written in the form

$$\tilde{\nabla}^2\tilde{\mathbf{A}} = -\mu_0\tilde{\mathbf{J}}. \tag{6.69}$$

It is also possible to write the curl equations in four-vector form but I had better stop here, you have probably had enough of mathematics.

I have introduced four-vectors merely to show the elegance of the notation. I have never come across a paper concerned with any practical applications (however remote) that would have employed the four-vector notation. Isn't it then a waste of time to talk about four-vectors? Perhaps. But having decided to include relativity, a few words on four-vectors seemed the right way of concluding the chapter.

Examples 6

6.1. The power density on Earth due to the radiation of the sun is $1{\cdot}35\ \mathrm{kw\ m^{-2}}$. Determine the amount of mass lost per second by the sun. The distance between the sun and the Earth is about $1{\cdot}5 \times 10^{11}$ m.

4.2. Prove that the momentum of light is equal to $p_l = W/c$ from the following argument. A carriage of mass M capable of moving without friction (Fig. 6.3) is isolated from its surroundings. At time $t = 0$ a light pulse is emitted by one of the inner walls. If light carries momentum p_l then the carriage recoils with velocity v obtainable from the relationship $v = p_l/M$. The carriage keeps on moving with this velocity until the light pulse hits the opposite wall (at $\Delta t = l/c$) where it delivers the same momentum and brings the carriage to a halt.

Determine the momentum of light from the condition that the centre of mass stays at the same place (don't forget the transfer of mass from one end of the carriage to the other).

FIG. 6.3. Frictionless carriage.

6.3. Assuming that the momentum of light is equal to W/c calculate the average pressure of light incident perpendicularly upon a completely absorbing body. Compare your result with that of Example 5.6.

6.4. If a body of mass m_0 acquires a speed v what is the increase (i) in its mass, (ii) in its energy?

Show that this excess energy reduces to $\frac{1}{2}m_0v^2$ when $v \ll c$.

6.5. By what voltage should an electron be accelerated in order to double its mass? At what velocity will it move then?

6.6. Assume that x' and t' are linear functions of x and t, and $y = y'$, $z = z'$. Assuming further that the expression (6.9b) is invariant derive the Lorentz transformation formulae eqns (6.10a) and (6.10b).

6.7. A body is moving with a speed u' in the frame S' which itself moves with a speed v with respect to S (v and u' being in the same direction).

Determine the body's speed as measured in S. What is the value of u when $u' = v = \frac{1}{2}c$?

6.8. Two infinite plates are placed at a distance d from each other and have surface charge densities ρ_s and $-\rho_s$ respectively (Fig. 6.4).

(i) Determine the electric field in the stationary S frame.

(ii) Determine the electric and magnetic fields, currents and surface charges in S' moving with a velocity v with respect to S.

(iii) Explain physically the results when $v \ll c$.

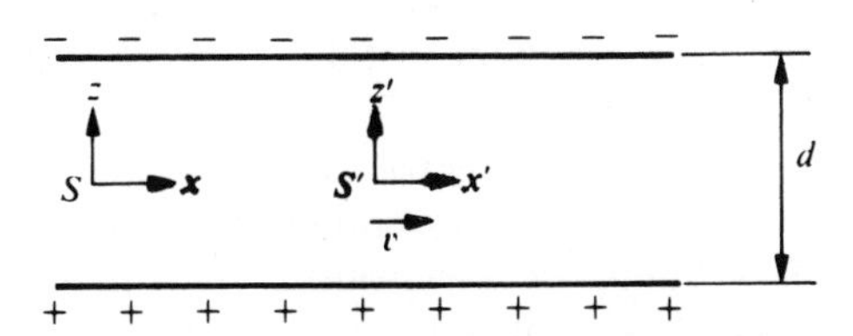

FIG. 6.4. Two parallel plates oppositely charged.

Epilogue

Natura simplex et fecunda,
The motto of Fresnel's prize winning essay.

Seek simplicity and distrust it,

A. N. WHITEHEAD

As I mentioned at the beginning electromagnetic theory has just the right amount of complexity for an engineer. The basic concepts (in contrast to quantum mechanics) are simple enough for anyone to grasp, at the same time the arising mathematical problems are complicated enough to present a challenge at each and every level of mathematical competence. We know the equations but we cannot solve them in their general form. We have to introduce simplifications and rely on physical intuition.

In the nineteenth century, in the heroic age of science, the faith in the simplicity of Nature was unwavering. Twentieth-century scientists are more sceptical but those working in electromagnetic theory still nurture that belief. As you may have noticed I always prefer a simple physical argument to a lengthy mathematical derivation. I would though like to give a warning. Just because something is simple it is not necessarily true. I talk from experience. It happened to me twice that I rushed into print with new theories yielding new physical pictures only to find a few years later that the results were valid in a much narrower range than I claimed.

Before I finish I have to say a few words about the Cinderella of this course, computers. Unfortunately, the application of numerical analysis has become so widespread that even a brief exposition of the various techniques used would require a separate course. To the question 'are we better off with computers?' the answer is an unconditional yes. First of all computers are useful in engineering design. Once the specifications (for an antenna for example) are given we may rely on the computer to find for us an optimum solution. Moreover, in many cases computer solutions are equivalent to experiments. Provided we can be sure of the accuracy of our results a solution for a set of parameters is as good as if we had actually performed the experiment. Finding agreement between 'theory' and 'experiment' means then the setting up of an approximate analytical theory for reproducing the computer results.

Finally, I have to admit to a certain bias. I am not indifferent towards electromagnetic theory; I have developed a certain liking for it. Other theories might go deeper but none of them (to my mind anyway) can explain a greater variety of phenomena. I hope that some of my enthusiasm has rubbed on to you.

Appendix

A.1. Vector identities

$$\nabla(\phi\psi) = \phi\,\nabla\psi + \psi\,\nabla\phi. \tag{A.1}$$

$$\nabla\,.\,(\phi\mathbf{E}) = \phi\nabla\,.\,\mathbf{E} + \mathbf{E}\,.\,\nabla\phi. \tag{A.2}$$

$$\nabla\times(\phi\mathbf{E}) = \phi\nabla\times\mathbf{E} - \mathbf{E}\times\nabla\phi. \tag{A.3}$$

$$\nabla\,.\,(\mathbf{E}\times\mathbf{H}) = \mathbf{H}\,.\,(\nabla\times\mathbf{E}) - \mathbf{E}\,.\,(\nabla\times\mathbf{H}). \tag{A.4}$$

$$\nabla(\mathbf{E}\,.\,\mathbf{D}) = (\mathbf{E}\,.\,\nabla)\mathbf{D} + (\mathbf{D}\,.\,\nabla)\mathbf{E} + \mathbf{E}\times(\nabla\times\mathbf{D}) + \mathbf{D}\times(\nabla\times\mathbf{E}). \tag{A.5}$$

$$\nabla\times(\nabla\times\mathbf{E}) = \nabla\nabla\,.\,\mathbf{E} - \nabla^2\mathbf{E}. \tag{A.6}$$

$$\nabla\times(\mathbf{E}\times\mathbf{H}) = \mathbf{E}\,\nabla\,.\,\mathbf{H} - \mathbf{H}\,\nabla\,.\,\mathbf{E} + (\mathbf{H}\,.\,\nabla)\mathbf{E} - (\mathbf{E}\,.\,\nabla)\mathbf{H}. \tag{A.7}$$

$$\begin{aligned}
(\mathbf{E}\,.\,\nabla)\mathbf{D} &= \left(E_x\frac{\partial}{\partial x} + E_y\frac{\partial}{\partial y} + E_z\frac{\partial}{\partial z}\right)\mathbf{D} \\
&= E_x\frac{\partial D_x}{\partial x}\mathbf{i}_x + E_x\frac{\partial D_y}{\partial x}\mathbf{i}_y + E_x\frac{\partial D_z}{\partial x}\mathbf{i}_z \\
&\quad + E_y\frac{\partial D_x}{\partial y}\mathbf{i}_x + E_y\frac{\partial D_y}{\partial y}\mathbf{i}_y + E_y\frac{\partial D_z}{\partial y}\mathbf{i}_z \\
&\quad + E_z\frac{\partial D_x}{\partial z}\mathbf{i}_x + E_z\frac{\partial D_y}{\partial z}\mathbf{i}_y + E_z\frac{\partial D_z}{\partial z}\mathbf{i}_z \\
&= \begin{bmatrix} \dfrac{\partial D_x}{\partial x}, & \dfrac{\partial D_x}{\partial y}, & \dfrac{\partial D_x}{\partial z} \\ \dfrac{\partial D_y}{\partial x}, & \dfrac{\partial D_y}{\partial y}, & \dfrac{\partial D_y}{\partial z} \\ \dfrac{\partial D_z}{\partial x}, & \dfrac{\partial D_z}{\partial y}, & \dfrac{\partial D_z}{\partial z} \end{bmatrix} \begin{bmatrix} E_x \\ E_y \\ E_z \end{bmatrix} \\
&= \frac{\partial\mathbf{D}}{\partial\mathbf{r}}\,.\,\mathbf{E}.
\end{aligned} \tag{A.8}$$

A.2. Vector operators in curvilinear coordinate systems

$$\nabla\times\mathbf{E} = \frac{1}{e_1e_2e_3}\begin{vmatrix} e_1\mathbf{i}_1 & e_2\mathbf{i}_2 & e_3\mathbf{i}_3 \\ \dfrac{\partial}{\partial x_1} & \dfrac{\partial}{\partial x_2} & \dfrac{\partial}{\partial x_3} \\ e_1E_1 & e_2E_2 & e_3E_3 \end{vmatrix}. \tag{A.9}$$

$$\nabla \cdot \mathbf{D}=\frac{1}{e_1e_2e_3}\left\{\frac{\partial}{\partial x_1}(e_2e_3D_1)+\frac{\partial}{\partial x_2}(e_3e_1D_2)+\frac{\partial}{\partial x_3}(e_1e_2D_3)\right\}. \tag{A.10}$$

$$\nabla\phi=\frac{1}{e_1}\frac{\partial\phi}{\partial x_1}\mathbf{i}_1+\frac{1}{e_2}\frac{\partial\phi}{\partial x_2}\mathbf{i}_2+\frac{1}{e_3}\frac{\partial\phi}{\partial x_3}\mathbf{i}_3. \tag{A.11}$$

$$\nabla^2\phi=\frac{1}{e_1e_2e_3}\left\{\frac{\partial}{\partial x_1}\left(\frac{e_2e_3}{e_1}\frac{\partial\phi}{\partial x_1}\right)+\frac{\partial}{\partial x_2}\left(\frac{e_3e_1}{e_2}\frac{\partial\phi}{\partial x_2}\right)+\frac{\partial}{\partial x_3}\left(\frac{e_1e_2}{e_3}\frac{\partial\phi}{\partial x_3}\right)\right\} \tag{A.12}$$

Rectangular coordinates

$$x_1=x, \quad x_2=y, \quad x_3=z; \quad e_1=e_2=e_3=1. \tag{A.13}$$

Cylindrical coordinates

$$x_1=r, \quad x_2=\varphi, \quad x_3=z; \quad e_1=1, \quad e_2=r, \quad e_3=1. \tag{A.14}$$

Spherical coordinates

$$x_1=r, \quad x_2=\theta, \quad x_3=\varphi; \quad e_1=1, \quad e_2=r, \quad e_3=r\sin\theta. \tag{A.15}$$

A.3. Physical constants

Mass of electron m	$9{\cdot}11\times10^{-31}$ kg
Charge of electron e	$1{\cdot}60\times10^{-19}$
Velocity of light c	$3{\cdot}00\times10^{8}$ m s^{-1}
Vacuum permeability μ	$4\pi\times10^{-7}$ V s A^{-1} m^{-1} (or H m^{-1})
Vacuum relative permittivity ε_0	$8{\cdot}85\times10^{-12}$ A s V^{-1} m^{-1} (or F m^{-1})

Translations of quotations

Si on parle trois langues on est trilingue,
Si on parle deux langues on est bilingue,
Si on ne parle qu'une langue on est un maudit Anglo-Saxon.

French Canadian joke

Writing from Brasenose College where the average number of languages a Fellow can read is well above three, the above indictment may appear less damning. It needs to be admitted however that some representatives of the English-speaking world might not be quite as assiduous in their studies of alien tongues. Hence translations will be given below of the Latin, German, Russian, and Danish texts (French will be assumed to cause no difficulties, not even the 17th-century French of a Dutchman).

To the question, 'What is Maxwell's theory?' I know of no shorter or more definite answer than the following: Maxwell's theory is Maxwell's system of equations.

HEINRICH HERTZ *Electric waves*, Macmillan London 1893 (translated by D. E. Jones)

The opposite ends of the galvanic battery were joined by a metallic wire, which, for shortness sake, we shall call the uniting conductor, or the uniting wire. To the effect which takes place in this conductor and in the surrounding space, we shall give the name of the conflict of electricity.

Let the straight part of this wire be placed horizontally above the magnetic needle, properly suspended, and parallel to it. If necessary, the uniting wire is bent so as to assume a proper position for the experiment. Things being in this state, the needle will be moved, and the end of it next the negative side of the battery will go westward.

The effect of the uniting wire passes to the needle through glass, metals, wood, water, resin, stoneware, stones; for it is not taken away by interposing plates of glass, metal or wood. Even glass, metal, and wood, interposed at once, do not destroy, and indeed scarcely diminish the effect. We found the effects unchanged when the needle was included in a brass box filled with water. It is needless to observe that the transmission of effects through all these matters has never before been observed in electricity and galvanism. The effects, therefore, which take place in the conflict of electricity are very different from the effects of either of the electricities.

Experiments on the Effect of a Current of Electricity on the Magnetic Needle. By John Christian Oersted, Knight of the Order of Danneborg, Professor of Natural Philosophy, and Secretary to the Royal Society of Copenhagen. Translated from a printed account drawn up in Latin by the author, and transmitted by him to the Editor of the Annals of Philosophy. Thomson's *Annals of Philosophy* **XVI** pp. 273–276 1820

The author himself has for a long time assumed a system in which all the internal mechanisms in solids like electricity, magnetism, heat and light as well as chemical affinities and separations may be attributed to the same fundamental forces. This system which he has proposed earlier in separate dissertations was more fully developed in his Ansichten der chemischen Naturgesetze published in 1812, and already at that time he came to the conclusion that magnetism may be produced by electric forces in their most concentrated state. He believed for a long time that to confirm this idea with experiments would be more difficult than actually proved to be the case. For this reason he devoted himself to other investigations until, in connection with his lectures on electricity, galvanism and magnetism in the spring of 1820, he decided once more to pursue this idea. He discovered then that the conductor which connects the opposite poles of the galvanic chain, and in which the effect detectable by any electrometer, has disappeared, has a powerful side effect which sets the magnetic needle into motion.

H. C. OERSTED *Communication about the discovery of electromagnetism, Bulletins of the Scientific Society* 1820–21 p. 12

We must mention here a rather unique piece of apparatus closely related to the equipment described above, for which we have to thank Professor Weber. Already last year he led a double wire connection over the roofs of the city between the Physics Laboratorium and the Observatory, and which is now extended from the Observatory to the magnetic measuring laboratory. This way a large galvanic chain is built in which the galvanic current (including the multiplicators at the endpoints) runs for a distance of ninethousand feet . . .

. . . The ease and reliability with which the direction of the currents by the Commutator, and the consequent motion of the needle could be controlled gave rise in the last year to experiments concerned with the transmission of telegraph signals. We were completely successful both with whole words and with short phrases. There is no doubt that in a similar manner it should be possible to establish direct telegraphic connection between places a considerable number of miles apart; but this cannot be the place where ideas about this matter could be further discussed.

CARL FRIEDRICH GAUSS *Magnetische Beobachtungen, Göttingische gelehrte Anzeigen* 1834 August 9

The investigation was arranged according to the following plan:—In the first place, regular progressive waves were to be produced in a straight, stretched wire by means of corresponding rapid oscillations of a primary conductor. Next, a secondary conductor was to be exposed simultaneously to the influence of the waves propagated through the wire and to the direct action of the primary conductor propagated through the air; and thus both actions were to be made to interfere. Finally, such interferences were to be produced at different distances from the primary circuit, so as to find out whether the oscillations of the electric force at great distances would or would not exhibit a retardation of phase, as compared with the oscillations in the neighbourhood of the primary circuit. This plan has proved to be in all respects practicable. The experiments carried out in accordance with it have shown that the inductive action is undoubtedly propagated with a finite velocity.

HEINRICH HERTZ *On the finite velocity of propagation of electromagnetic actions in Electric waves* Macmillan London 1893 (translated by D. E. Jones)

Certainly it is a fascinating idea that the processes in air which we have been investigating represent to us on a million-fold larger scale the same processes which go on in the neighbourhood of a Fresnel mirror or between the glass plates used for exhibiting Newton's rings.

HEINRICH HERTZ *On electromagnetic waves in air and their reflection* in *Electric waves* Macmillan London 1893 (translated by D. E. Jones)

I have taken up the experiments in the simple form suggested by Maxwell . . . .
These forces exerted by the light agree within experimental error with the ponderomotive forces of radiation calculated by Maxwell and Bartoli.
Moscow, Physics Laboratory of the University.

P. LEBEDEW *Investigation of forces exerted by light, Ann. Phys.* **4**, **6** 1901

Proposition I
Light propagates and spreads out not only directly, by refraction and by reflection but also in a fourth way, by diffraction.

Brilliant sunshine is introduced into an otherwise completely dark chamber by an extremely small aperture AB. The light spreads out following the cone ACDB which is visible if the air contains a lot of dust or some smoke is produced. An opaque body EF is inserted into this cone at a large distance from the aperture AB, so that at least one edge of the opaque body is illuminated.

Furthermore, certain divided and distributed tracts or series of coloured light may be observed over the brightly lit CM and ND parts of the light base. The light is pure and natural in the middle of any of these series but there are some colours at the edges, blue nearer to the shadow MN and red in the other direction.

Proposition XXII
The surface of an illuminated body may be made darker by the reception of some new light.
This proposition appears to be paradoxical because it shows a highly improbable state of affairs. The function of light is to illuminate the opaque body upon which it is incident and not to make it darker. The phenomenon is however quite well proven and may be easily demonstrated by some experiments though as far as I know no one has so far done them. I shall now proceed briefly to describe them.

F. M. GRIMALDI *Physico-mathematical treatise on light, colours, the rainbow and other related topics* Bologna 1665

As we saw however, closer observation shows an extremely weak scattering of light that takes place at the very edge of the screen. This may be seen on photograph 13 displaying the light rays moving along the surface of the photographic plate upon which the screen PQ stands. Since the scattered light is a few million times weaker than the direct light the sections *c* and *d* of the plate (where they are displayed) had to be exposed a few million times longer than the middle sections *a* and *b*. For this reason the photograph may only be obtained artificially by superposing three photographs of which the middle one is taken separately from those on the two sides obtained by the above described method. From the point of view of Huygens' principle this phenomenon may be looked upon as light emisson from secondary illuminated points located on the very edge of the screen.

A. KALASHNIKOW *Gouy–Sommerfeld diffraction, Journal of the Russian Physical and Chemical Society* **44** 1912

Nonetheless it is interesting to see that Young's theory concerned with the illumination of the edge of the screen is not confirmed here just by coincidence. We shall actually show that the function . . . may be split by a transformation into an (in the sense of the geometrical optics) incident light wave and into a diffraction wave originating at the edge of the screen. We may thus regard the diffraction wave as obeying the elementary law that each point of the aperture edge emits an unsymmetric spherical wave, determined solely by the relative geometrical positions of the light source and of the relevant aperture element.

A. RUBINOWITZ *Annalen der Physik* **53** 1917

. . . if we wish to talk of a reflection of the incident light as Thomas Young did, then the 'kind of reflection' is very specialized and must be defined precisely.

The question arises whether and in what way Young's interpretation may be extended to arbitrary diffraction screens.

ARNOLD SOMMERFELD *Lectures on Theoretical Physics Vol. IV, Optics* Academic Press 1964

Take, for example, the reciprocal electrodynamic action of a magnet and a conductor. The observable phenomenon here depends only on the relative motion of the conductor and the magnet, whereas the customary view draws a sharp distinction between the two cases in which either one or the other of these bodies is in motion. For if the magnet is in motion and the conductor at rest, there arises in the neighbourhood of the magnet an electric field with a certain definite energy, producing a current at the places where parts of the conductor are situated. But if the magnet is stationary and the conductor in motion, no electric field arises in the neighbourhood of the magnet. In the conductor, however, we find an electromotive force, to which in itself there is no corresponding energy, but which gives rise—assuming equality of relative motion in the two cases discussed—to electric currents of the same path and intensity as those produced by the electric forces in the former case.

Examples of this sort, together with the unsuccessful attempts to discover any motion of the earth relatively to the 'light medium', suggest that the phenomena of electrodynamics as well as of mechanics possess no properties corresponding to the idea of absolute rest.

The introduction of a 'luminiferous ether' will prove to be superfluous inasmuch as the view here to be developed will not require an 'absolutely stationary space' provided with special properties.

ALBERT EINSTEIN *On the electrodynamics of moving bodies Annalen der Physik* **17** 1905

If a body gives off the energy L in the form of radiation, its mass diminishes by L/c^2. The fact that the energy withdrawn from the body becomes energy of radiation evidently makes no difference, so that we are led to the more general conclusion that

The mass of a body is a measure of its energy-content; if the energy changes by L, the mass changes in the same sense by $L/9 \times 10^{20}$, the energy being measured in ergs, and the mass in grammes.

It is not impossible that with bodies whose energy-content is variable to a high degree (e.g. with radium salts) the theory may be successfully put to the test.

If the theory corresponds to the facts, radiation conveys inertia between the emitting and absorbing bodies.

ALBERT EINSTEIN *Does the inertia of a body depend upon its energy content? Annalen der Physik* **18** 1905

Answers to examples

Chapter 2

(1) The force on the charge in the upper right hand corner is

$$\mathbf{F} = \frac{q^2}{16\pi\varepsilon_0 a^2}(4-\sqrt{2})(\mathbf{i}_x + \mathbf{i}_y)$$

with similar equations for the other charges.

(2) $E = \dfrac{qr}{4\pi\varepsilon_0 a^3},\ r < a,\ E = \dfrac{q}{4\pi\varepsilon_0 r^2},\ r > a.$

(3) $E = \dfrac{\rho_l a}{2\varepsilon_0} \dfrac{|x|}{(x^2+a^2)^{\frac{3}{2}}}$ where x is the distance along the axis measured from the centre of the ring.

(4) $E_R = \dfrac{\rho_l}{4\pi\varepsilon_0 R}\left[1 - \dfrac{z}{(z^2+R^2)^{\frac{1}{2}}}\right],\ E_z = \dfrac{\rho_l}{4\pi\varepsilon_0 (R^2+z^2)^{\frac{1}{2}}}.$

(5) $E_R = \dfrac{3qd^2}{2\pi\varepsilon_0 r^4}(3\cos^2\theta - 1),\quad E_\theta = \dfrac{3q^2 d}{2\pi\varepsilon_0 r^4}\sin 2\theta.$

(6) (i) $E_\varphi = 0$ at $R = a$, (ii) $\dfrac{dy}{dx} = \dfrac{2a^2xy}{(x^2+y^2)^2 + a^2(x^2-y^2)}.$

(8) $\left(x + \dfrac{d}{p^2-1}\right)^2 + y^2 + z^2 = \left(\dfrac{dp}{p^2-1}\right)^2.$

(9) $F = \dfrac{q^2}{4\pi\varepsilon_0} \dfrac{Ra}{(R^2-a^2)^2}$ where R is the radius of the sphere.

(10) $C = \dfrac{\pi\varepsilon_0}{\ln[4d_1d_2/D(d_1^2+4d_2^2)^{\frac{1}{2}}]}.$

(12)

$$E_{\theta 2} = E_0 \frac{3a\sin\theta}{\varepsilon_r + 2} = E_{\theta 1},$$

$$E_{r2} = -E_0 \frac{3\varepsilon_r \cos\theta}{\varepsilon_r + 2} = \varepsilon_r E_{r1},$$

where the subscripts 1 and 2 refer to quantities inside and outside the dielectric sphere respectively.

(13) $\dfrac{dC/C}{d\alpha/\alpha}=\dfrac{\alpha(\varepsilon_r-1)}{1+\alpha(\varepsilon_r-1)}.$

(14) (i) $y^2=4v^2(x+v^2)$; $y^2=4u^2(u^2-x^2)$.

(ii) $x^2+\left(y+\dfrac{1}{2v}\right)^2=\left(\dfrac{1}{2v}\right)^2$; $\left(x-\dfrac{1}{2u}\right)^2+y^2=\left(\dfrac{1}{2u}\right)^2$.

(iii) $x^2+y^2=\exp(2u)$; $y=x\tan v$.

(15) $W_{\max}=E_b^2\pi\varepsilon la^2\ln b/a$; $W_{\max}=4{\cdot}31$ ws.

(16) $F=\dfrac{V^2}{2}\dfrac{t}{d}\varepsilon_0(\varepsilon_r-1)$ where t is the width of the capacitor plate.

(17) $V=\left(\dfrac{8kd^3}{27\varepsilon_0 S}\right)^{\frac{1}{2}}$

(18) (i) $T=-\dfrac{A}{2}V^2$, (ii) $T=-\dfrac{A}{2}\dfrac{\pi^2V^2}{\theta^2}$ where $A=(n-\tfrac{1}{2})\dfrac{r^2\varepsilon}{t}$, breakdown and/or scattered capacitance.

Chapter 3

(1) $H_\varphi=0,\quad r<a,\quad H_\varphi=\dfrac{(r^2-a^2)J_0}{2r},\quad a<r<b.$

$H_\varphi=\dfrac{(b^2-a^2)J_0}{2r},\quad r>b.$

(2) $p(R)=\dfrac{\mu_0J_0^2}{4}\left(\dfrac{D^2}{4}-R^2\right)$ where D is the diameter of the column and R is an arbitrary radius. $p(2{\cdot}5)=9{\cdot}55\ \mathrm{Nm^{-2}}$, $p(0)=12{\cdot}57\ \mathrm{Nm^{-2}}$.

(4) $V_1=\dfrac{d_1\sigma_2}{d_1\sigma_2+d_2\sigma_1}V,\quad V_2=\dfrac{d_2\sigma_1}{d_1\sigma_2+d_2\sigma_1}V,\quad \rho_s=\dfrac{\varepsilon_2\sigma_1-\varepsilon_1\sigma_2}{d_1\sigma_2+d_2\sigma_1}V.$

(5) $V=\tfrac{1}{4}k^{\frac{2}{3}}(3x)^{\frac{4}{3}}$ where $k=\dfrac{J}{\varepsilon_0}\left(-\dfrac{m}{2e}\right)^{\frac{1}{2}}.$

(6) $\dfrac{I}{b\pi}\tan^{-1}\dfrac{b}{2a}.$

(7) (iii) $H=\dfrac{In}{2\pi a}\tan\dfrac{\pi}{n}.$

(9) $A_\varphi=6\times10^{-9}\ \mathrm{Vsm^{-1}}$.

(11) $H_R = \frac{\mu_0 I}{2\pi} \frac{z}{R[(R+a)^2+z^2]^{\frac{1}{2}}} \left[\frac{R^2+a^2+z^2}{(R-a)^2+z^2} E - K \right],$

$H_z = -\frac{\mu_0 I}{2\pi} \frac{1}{[(R+a)^2+z^2]^{\frac{1}{2}}} \left[\frac{R^2+z^2-a^2}{(R-a)^2+z^2} E - K \right],$

$H_R = 4{\cdot}54 \times 10^{-8}\ \text{Am}^{-1}, \quad H_z = 1{\cdot}09 \times 10^{-7}\ \text{Am}^{-1}.$

(12) Yes.

(13) 1·2T, 5·48A, 1·35T.

(14) $b = 32$ mm, radius of copper wire $= 0{\cdot}302$ mm, $N = 2770$, $h_1 =$ 10 mm, $h_2 = 6{\cdot}25$ mm.

Chapter 4

(1) $E = 99{\cdot}2$ mV.

(2) $E = \frac{\sqrt{3}}{4\pi r^2} \mu_0 abl\omega I_0, \quad r = \left[d^2 + \left(\frac{a}{2} \right)^2 \right]^{\frac{1}{2}}, \quad E = 6{\cdot}45\ \mu\text{V}.$

(3) $I_l = \frac{\text{j}\omega T}{1+\text{j}\omega T} H_0, \quad H_1 = \frac{H_0}{1+\text{j}\omega T}, \quad P_l = \frac{\pi R}{\delta\sigma} \frac{\omega^2 T^2 H_0^2}{1+\omega^2 T^2}$

where $T = \frac{\mu_0 R \delta\sigma}{2}$.

(5) An approximation introduced earlier that $a_2 \ll a_1$.

(6) 105·2 μH.

(7) Force repulsive. Yes, Al preferable.

(8) (i) $\frac{\text{d}v}{\text{d}t} + Av = g, \quad A = \frac{B_r^2 \sigma}{d}$ where d is density.

(ii) $v = \frac{g}{A}(1 - \text{e}^{-At}), \quad s = \frac{g}{A}\left[t - \frac{1}{A}(1 - \text{e}^{-At}) \right].$

(iii) $v = \frac{g}{A}$, (iv) Al falls faster.

(9) $N = 1000, \quad I_m = 90$ mA.

(10) Gives, of course, the same as eqn(4-90).

(11) $E = \frac{2\mu_0 Ikb\omega \sin \omega t}{k^2 - \cos^2 \omega t}$ where $k = \frac{a^2 + 4d^2}{4ad}$.

(12) $E = \frac{4\mu_0 abvI}{\pi(4d^2 - a^2)}$.

(13) $\theta=45°$; $I_2=0$, $T=0{\cdot}4$ Nm; $I_2=2$A, $T=1{\cdot}81$ Nm.

$\theta=135°$; $I_2=0$, $T=-0{\cdot}4$ Nm; $I_2=2$A, $T=1{\cdot}01$ Nm.

Chapter 5

(1) $P_d=\dfrac{I^2}{2\sigma\pi^{1/2}S_0^{3/2}}$, power in = resistive loss.

(3) (i) $H=\sqrt{\dfrac{\varepsilon_0}{\mu_0}}\,[-E_2\sin(\omega t-\beta_0 z)\mathbf{i}_x+E_1\cos(\omega t-\beta_0 z)\mathbf{i}_y]$.

(ii) $\left(\dfrac{E_x}{E_1}\right)^2+\left(\dfrac{E_y}{E_2}\right)^2=1.$

(4) About 10 mm. Yes, because many of the conductors are thicker than 20 mm.

(5) (ii) $\dfrac{1}{2}\left(\dfrac{\omega\mu}{2\sigma}\right)^{\frac{1}{2}}|H_y|^2$.

(8) $H_x=\mathrm{j}\dfrac{\beta}{(\beta^2-\beta_0^2)}\dfrac{\partial H_z}{\partial x}$, $\quad H_y=\mathrm{j}\dfrac{\beta}{(\beta^2-\beta_0^2)}\dfrac{\partial H_z}{\partial y}$.

$E_x=-\dfrac{1}{\mathrm{j}\omega\varepsilon_0}\dfrac{\beta_0^2}{(\beta^2-\beta_0^2)}\dfrac{\partial H_z}{\partial y}$, $\quad E_y=\dfrac{1}{\mathrm{j}\omega\varepsilon_0}\dfrac{\beta_0^2}{(\beta^2-\beta_0^2)}\dfrac{\partial H_z}{\partial x}$.

(9) (iii) $\beta_y=\dfrac{m\pi}{a}$ $\quad\beta_x=\dfrac{n\pi}{b}$.

(iv) $f_c=\dfrac{c}{2}\sqrt{\left(\dfrac{6}{a}\right)^2+\left(\dfrac{7}{b}\right)^2}$.

(13) (iv) 0·082 db/m, (v) 0·069 db/m, $f=14{\cdot}5$ GHz.

(14) (i) $f_c=\dfrac{3{\cdot}832}{2\pi}\dfrac{c}{R_0}$.

(16) 234 km.

(17) (ii) $f_r=7{\cdot}72$ GHz.

(18) $n\cong 502$.

(21) $Q=5{\cdot}16\times10^6$.

(22) $\varepsilon_r=1{\cdot}51$.

(23) $n=16700$, $\Delta f\cong\dfrac{f}{n}=30$ GHz.

(24) Because E and H behave differently at a boundary.

(25) (i) $\beta_{m2}d = \frac{\pi}{2}$, $Z_2 = (Z_1 Z_3)^{\frac{1}{2}}$; $\beta_{m2}d = \pi$, $Z_1 = Z_3$.

(ii) Matching in general, e.g. photographic lenses.

(iii) $\Gamma_{max} = 0{\cdot}882$.

(27) (ii) $H_\phi = \dfrac{jI_m}{2\pi r}\exp(-j\beta_0 r_0)\dfrac{\cos\left(\frac{\pi}{2}\cos\theta\right)}{\sin\theta}$, $E_\theta = Z_0 H_\phi$.

(iv) $P_d = \dfrac{1}{2\pi}Z_0 I_m^2 \displaystyle\int_0^{\pi/2} \dfrac{\cos^2[(\pi/2)\cos\theta]}{\sin\theta}\,d\theta$.

(v) $R = 73$ ohm.

(vi) $G = 1{\cdot}65$.

(28) The far side of the lens is plane whereas the near side obeys the equation (in spherical coordinates, r, θ)

$$r = \frac{(n-1)f}{n\cos\theta - 1}$$

where $n = \varepsilon_r^{\frac{1}{2}}$ and f is the focal distance (distance of the point source from the nearest point on the lens).

(29) (i) $g(\theta, 0) = \left|\dfrac{\sin u}{u}\right|$, (ii) $g(\theta, 90°) = \left(\dfrac{\pi}{2}\right)^2 \left|\dfrac{\cos v}{v^2 - (\pi/2)^2}\right|$,

both radiation patterns normalized to unity.

(iii) $G = 0{\cdot}81\dfrac{4\pi ab}{\lambda^2}$, (iv) smaller by the factor $1{\cdot}24$.

(30) (i) $g(\theta, 0) = \left|\dfrac{\sin w}{w}\right|$ where $w = \dfrac{\beta_0 a}{2}\sin\theta - \alpha$.

(ii) $\theta_{max} = \dfrac{2\alpha}{\beta_0 a}$.

Chapter 6

(1) $4{\cdot}24\ 10^9\,\mathrm{kg\,s^{-1}}$.

(3) $p_{average} = \frac{1}{2}\mu H^2$ where H is the magnetic field of the incident wave. Result is different from that of Example 5.6 because the physical picture is different. In the present case the wave instantly loses all its momentum when it falls upon the absorbing body.

(4) $\Delta m = m_0\left[\left(1-\frac{v^2}{c^2}\right)^{-\frac{1}{2}}-1\right].$

$\Delta W = m_0 c^2\left[\left(1-\frac{v^2}{c^2}\right)^{-\frac{1}{2}}-1\right].$

(5) $v = 0{\cdot}866c, \quad V = 570\ \text{kV}.$

(7) $u = \dfrac{u'+v}{1+\dfrac{vu'}{c^2}}, \quad u = \dfrac{4c}{5}.$

(8) (i) $E_z = \rho_s/\varepsilon_0$, (ii) $E'_z = \gamma E_z$, $B^1_y = \frac{\gamma v}{c^2}E_z$, $K'_x = -v\gamma\rho_s$, $\rho'_s = \gamma\rho_s$, (iii) the electric field and surface charge density will remain the same but in S' the surface charges appear to move in the opposite direction giving rise to a surface current and hence to a magnetic field.

Author Index
(for the quotations only)

General Index